Sustaining the Earth

Sustaining the Earth

CHOOSING CONSUMER PRODUCTS THAT ARE SAFE FOR YOU, YOUR FAMILY, AND THE EARTH

DEBRA DADD-REDALIA

HEARST BOOKS
NEW YORK

Permission to reprint "What can one brewer do to help the environment?" advertisement on pages 52–53 courtesy of Anheuser-Busch Companies.

Description of hybrid-natural ingredient by Philip Dickey of the Washington Toxics Coalition in *Green Alternatives* (May/June 1992). Reprinted with permission.

Discussion of socially responsible business on pages 190–191. Reprinted with permission by Doug Tomkin and Paul Hawken.

The Seal of Cotton is a registered servicemark/trademark of Cotton Incorporated.

Excerpt from *Green Marketing: Hope or Hoax?* by Ruth Loomis and Merv Wilkinson, New Society Publishers. Reprinted with permission.

It is the policy of William Morrow and Company, Inc., and its imprints and affiliates, recognizing the importance of preserving what has been written, to print the books we publish on acid-free paper, and we exert our best efforts to that end.

Library of Congress Cataloging-in-Publication Data
Dadd-Redalia, Debra.
Sustaining the earth : Choosing consumer products that are safe for you, your family, and the earth / Debra Dadd-Redalia.
p. cm.
Includes bibliographical references and index.
ISBN 0-688-12335-X
1. Environmental health. 2. Product safety. 3. Consumer education. I. Title.
RA565.D343 1994
363.7—dc20 93-49649
CIP

Printed in the United States of America

First Edition

1 2 3 4 5 6 7 8 9 10

BOOK DESIGN BY CLAIRE NAYLON VACCARO

For my father,
Robert Charles Dadd

Contents

PART TWO

Acknowledgments

I would be remiss if I did not acknowledge that the idea for this book was not entirely my own. It is the result of many discussions held between the summer of 1990 and the summer of 1991, during which I participated in the research and planning of a chain of environmental retail stores for WorldWise, Inc. Because part of my task was to create criteria for choosing products for these stores, I had the time and support to delve deeply into the issues surrounding choosing products that had legitimate environmental claims, which led me to sustainability as a guiding principle. Though WorldWise, Inc., ultimately decided to serve the wholesale rather than retail market, going through this process made two things clear: We need to think more deeply when assessing products for their environmental benefits, and consumers need to know more about how to evaluate products they find on the shelves. I want to particularly thank Aaron Lamstein for many lively hours of conversation as he challenged and probed my emerging ideas.

My editor, Harriet Bell, made sure that I condensed a very complex subject into a tool that is truly useful. As the best editors do, she guided me to write what I really wanted to say, and would accept nothing less. Thanks, Harriet, for your patience and commitment.

I also want to thank Martha Casselman, my literary agent, for her usual wonderful job, and my husband, Larry Redalia, for continuing to be so loving even when this book occupied all my time.

And, finally, I want to acknowledge everyone who is working to improve the environmental integrity of products—all the manufacturers, certification and educational organizations, consumer advocates, retailers, and consumers. To make this immense shift in the marketplace requires our mutual cooperative effort. Not one of us can do it alone.

Author's Note

When we buy products that claim to have social or environmental benefits, we do so trusting that the claims made by the manufacturers are true. And, in buying a book such as this, you do so trusting that what is printed in these pages is true as well.

To the best of my knowledge, at the time of printing, the information on these pages is accurate and what I believe to be correct, both in philosophy and in fact. I have not personally checked out every product and catalog mentioned in this book. Some of these companies I know personally, but not all. No independent testing was done by me. Like any other consumer, I made product recommendations from written materials supplied by manufacturers or retailers, or from personal conversations with people who make and sell these products.

Specific businesses are mentioned as examples, not recommendations. No advertising or promotional fees were accepted, and the inclusion of any product, business, book, publication, or organization here is not an endorsement. If I occasionally mention that I use a product personally, this indicates only my preference; that item is not necessarily superior to other similar products.

To be a responsible consumer, you must make your own choices and do your own homework. Be as wary of blindly accepting the information presented here as you would of assuming that all products on the market are safe. Despite heroic efforts on my part to ensure accuracy, I cannot guarantee everything to be absolutely correct, though I have worked hard toward that goal. Specifically,

I can make no representations about, and therefore cannot be responsible for: incorrect or incomplete information regarding products; changes in product ingredients or production methods; substances or processes with toxicity or environmental destructiveness as yet undiscovered; the continued availability of any item; the effectiveness of any product or process listed; or any adverse health or environmental effects caused by any product or process listed.

Like any other type of business, companies that make or sell products with environmental and social benefits have been known to make fraudulent claims. It has been my experience, however, that companies with moral awareness high enough to make such products also generally are honest about their claims, particularly if they are small businesses. Build trusting relationships with the manufacturers of the products you use regularly by asking them for full-disclosure information. I've found that companies that are truly making sustainable improvements are more than happy to tell you just what they are doing.

Introduction

People cannot live apart from nature . . . And yet, people cannot live in nature without changing it. What we call nature is, in a sense, the sum of the changes made by all the various creatures and natural forces in their intricate actions and influences upon each other and upon their places. The making of these differences is the making of the world . . .

And so it can hardly be expected that humans would not change nature. Humans, like all other creatures, must make a difference; otherwise, they cannot live. But unlike other creatures, humans must make a choice as to the kind and scale of the difference they make. If they choose to make too small a difference, they diminish their humanity. If they choose to make too great a difference, they diminish nature, and narrow their subsequent choices; ultimately, they diminish or destroy themselves. Nature, then, is not only our source but also our limit and measure.

—Wendell Berry

Ten years ago, my first consumer guidebook was published. *Nontoxic & Natural* was a comprehensive list of more than six hundred mail-order catalogs and twelve hundred brand-name household products then available that did not contain toxic chemicals. In 1990 it was updated as *Nontoxic, Natural & Earthwise*, another comprehensive list of more than six hundred mail-order catalogs and two thousand brand-name household products that were safe for health and better for the environment.

My purpose in writing those two books was to introduce new philosophies for choosing products—first eliminating household toxics, and then buying products that are better for the environment—and to show that products existed in the marketplace that met these criteria by listing every such product and catalog I could find.

Sustaining the Earth is similar to my previous books in that it introduces an even newer philosophy by which we can evaluate everyday products—sustainability—but it is different in that it does not contain copious lists of products (although I do give a few examples for each product type). I didn't include every product I could find for a very specific reason: Product lists printed in books are a static snapshot of what was available when the author did the research. This book, in contrast, is designed to be a dynamic, interactive tool to be used to evaluate products for yourself, now and in the future. It is a basic primer that will give you background information on product assessment and guidelines for finding and choosing specific products, so that you can pick up virtually any product you find in a store or order from a catalog, knowledgeably assess its environmental pros and cons, and distinguish products with real environmental benefits from those that are nothing more than misleading "green" hype.

In the last few years, thousands of new products have been introduced into the marketplace that either claim to be or imply that they are in some way better for the environment. There has been a flurry of new environmental catalogs and greenstores opening around the country and environmental products can now be found on the shelves in supermarkets, discount warehouses, department stores, natural food stores, in mail-order catalogs, and almost everywhere else we shop. Marketing for the environment has become big business. One study by a market research firm cited in *Advertising Age* predicts that we'll be spending $8.8 billion a year on green products by 1995. This is not a passing fad—it's a trend in the right direction. Products that have environmental benefits are here to stay, and we need to learn about them.

In my work as a consumer advocate, one of the things that has troubled me over the last few years is that while many companies have been claiming their products are "good for the earth," there has been little definition of what exactly "good for the earth" means. We are all well aware that there are environmental problems, but what has not been so clear is what we can do that will really improve the situation. Recycling your bottles and cans is indisputably better for the earth than having a garbage truck take them to a landfill, but recycling alone is not the answer. Protecting the natural beauty of wilderness areas is also important, but having patches of pristine land separated from exploited areas occupied by humans will not save the environment either.

What one comes to, after asking over and over "What is good for the earth?," is that we must find a way to inhabit the earth in a way that is harmonious rather than destructive, and to do that, we must look to the earth—to nature itself—for the answers. Though it is not generally recognized in our modern times, we, the human species, are an integral part of the natural world—interdependent with other living things for our survival, and subject to all the same natural processes as other species. What is good for the earth is for us to live in harmony with its creatures and according to its ways, which in turn will be good for us.

This idea of living in harmony with nature has come to be known as sustainability, and is now being used worldwide as a guideline for developing new environmental programs. *Sustainable* means being able to keep going or able to endure; sustainability, then, is acting in such a way that life on earth endures on into the future, providing for the needs of all citizens and creatures while maintaining the natural functions, resources, and beauty of the planet. The most commonly used definition of sustainability is "to meet the needs of the present without compromising the ability of future generations to meet their own needs," which first appeared in the 1987 *Report of the World Commission on Environment and Development*. Sustainability is rooted in looking to the inherent workings of nature as a model, with the idea that the natural systems of the world do work in balance to perpetuate life, and by working in harmony with those natural systems, we can sustain our own lives.

The idea of a sustainable society includes both human needs and the needs of the earth, not one exclusive of the other. Instead of taking a stance that says "it's humans *or* the environment," as if humans and the environment were polar opposites, sustainability acknowledges that we are an integral part of the environment and shows us that the only way we can sustain ourselves is to sustain the earth.

The ideas for building sustainability are broad and need to be appropriately interpreted in all aspects of our lives and in all places on the earth. According to *Caring for the Earth: A Strategy for Sustainable Living* by The World Conservation Union, United Nations Environment Programme, and the World Wide Fund for Nature, the principles of a sustainable society include:

- Respecting and caring for the community of life;
- Improving the quality of human life;
- Conserving the Earth's vitality and diversity;
- Minimizing the depletion of nonrenewable resources;
- Keeping within the Earth's carrying capacity;
- Enabling communities to care for their own environments;
- Creating a global alliance.

Currently much of the way we live is unsustainable. We rely heavily on crude oil, for example, for everything from fueling our entire transportation system to making our polyester clothing, yet crude oil is a nonrenewable resource. Many experts agree that if we continue our current usage, we have a less than fifty-year supply of crude oil remaining and nature isn't making more anytime soon. And we're clear-cutting forests faster than they can regrow. Landfills are overflowing, species are disappearing, populations are exploding—if this trend continues, soon the earth will have nothing left to give.

Using sustainability to guide our actions, on the other hand, can lead us out of our environmental crisis to a world where there is a continuous renewal of life and resources. Already sustainable technologies such as renewable energy produced by sun and wind are being used to power American homes—both as a direct source of power and as a supplementary power source by local utility companies. Foresters are finding ways to selectively log trees and leave our domestic forests standing. Recyclable materials are being made into new products instead of sitting in landfills, whole habitats are being preserved, and in this country at least, our population is stabilizing. All of this is happening now and can happen more and more. Sustainability is a real option—we need only choose that path.

Creating a sustainable world is necessarily a global goal, for the sustainability of any one place is dependent on the sustainability of others. Pollution in air and water, for example, does not recognize man-made boundaries as it spreads through the environment and if we deplete our local resources, we have to pull resources from other places in order to survive. But sustainability cannot come about from global international protocols or government regulations—each of us must choose to live sustainably in our own lives, and work with our neighbors to live sustainably in our local communities. In time sustainable ways will become ingrained in our way of being as a society.

In practice, sustainability, by its very nature, requires a person (or community of people) and a place. Just as nature has different characteristics in the desert, the mountains, the tropics, so appropriate ways of living sustainably will be different in every place on earth. To live in harmony with nature requires knowing nature as it is in the place you live. The details of how I can best live sustainably here in a bay and fir forest near the Pacific Coast in a rural area of northern California are very different from how one might best live sustainably on a farm in Kansas, in an apartment in Manhattan, or in a village in a Third World country. Yet we all have the responsibility—inherent in the privilege of living on this earth—of sustaining the land on which we live and the life on which we depend.

I do not want to oversimplify this idea of sustainability. It is like a multifaceted jewel that will take many years and a multitude of minds, hearts, and hands to polish. We have much to learn. To live sustainably in our places requires fundamental changes in our lives and our society, but we must start somewhere. One of the easiest and most important things we can do right now as a first step toward sustaining the earth is to choose products that we use on a daily basis with sustainability in mind, for our combined consumer purchases make an enormous impact on the environment. To explore sustainability as it relates to choosing everyday products is the purpose of this book.

The title of this book is somewhat of an oxymoron—it is a guide for consumers about sustaining the earth, but to sustain the earth, we need also to rethink our role as consumers. A *consumer* is someone who consumes or someone who uses a commodity or service, and to *consume* is to destroy or expend by use, to use up, or to spend wastefully. Because the very purpose of our consumer culture is to destroy, expend, use up, and waste, being a consumer is not a sustainable activity.

The week after I wrote the preceding paragraph, an article appeared in my local newspaper called "What's enough stuff?" It pointed out how, as a consuming culture, we're hooked on stuff—not necessarily valuable stuff, just stuff. With every year that passes, it said, it takes more stuff to have enough stuff, and the frenzy is accelerating. The way people measure their success is in terms of material possessions. Piles of things, therefore, make consumers feel good about themselves; having only essential necessities makes them feel like a failure. Instead of accumulating only what they need, consumers are now moving into accumulating more than they need: five televisions, twenty pairs of shoes, a dozen kitchen appliances. A professor of retailing and marketing at our local junior college, who has held countless class discussions on the concept of having enough said, "We've come to the conclusion that . . . what is enough is when your money runs out."

While environmental degradation has been going on in varying degrees throughout the history of the world, there has been no other time that humans have taken so many resources from the earth and created so much waste as we have since the beginning of the industrial revolution, and particularly since the creation of a consumer culture. Being a consumer is not our natural, human state—our passion for consuming has been deliberately cultivated.

In the 1950s marketing consultant Victor Lebow wrote in the New York *Journal of Retailing*, "Our enormously productive economy demands that we make consumption a way of life, that we convert the buying and use of goods into rituals, that we seek our spiritual satisfaction in consumption . . . We need things consumed, burned up, worn out, replaced, and discarded at an ever-growing rate." Instead of buying items that meet our needs, the marketplace entices us to buy things we don't need. We consume because we've been conditioned by advertising to consume.

Many of us, myself included, were raised on this ethic—it's all we know. To be a good consumer was the lesson I learned from my parents. I was "born to shop." Quality time with my mother was spent shopping at the mall; she gave me my first credit card when I was sixteen. Going shopping was entertainment, a setting for social interaction, a way to see things that were new and different. I became so immersed in consumer culture that I made a career of it as a consumer advocate.

When I first became interested in living in a way that was in harmony with nature, I looked to indigenous cultures for inspiration. There is much we can learn from native peoples about living sustainably in our places, but we can't go back to living in such a primitive way. Even though I could see we needed to make a change in both how we choose products, and the way we live, a basic piece of the puzzle of what we need to do to live sustainably *now* didn't fall into place for me until I started reading Wendell Berry, particularly his *The Unsettling of America* and *Home Economics*. Given that his writing has been motivated "by a desire to make myself responsibly at home in this world and in my native and chosen place" (a very sustainable idea), I was surprised that I had never been drawn to read his books before. The reason, I believe, is that his books are promoted as being about small-scale agriculture and preservation of the family farm, but his insights on being a consumer and keeping a household give solid guidelines for creating a sustainable home and economy.

According to Mr. Berry, a consumer buys everything needed for survival, food, water, clothing, shelter, and as a consequence, consumers need an ever-increasing, steady supply of money in order to survive. As consumers we put our lives in the hands of those who sell the products and services we rely on because we do not produce any of our basic needs for ourselves. Our choices are dictated by advertising, salesmanship, and the amount of money we have. We think we gain convenience and leisure, but in the process we forfeit our creativity, our individuality, our ability to fend for ourselves, our human relationships, and our connection with the source of our sustenance. The whole idea of consumerism is disconnected from and therefore destructive of natural ecosystems. For a consumer domestic labor consists of buying things, putting them away, and throwing things away; if it can be managed, all domestic work is done by someone else. Consumers need instant gratification and often live on credit. It's a vicious trap that depletes financial, human, and environmental resources. In many ways our environmental crisis is fundamentally similar to our consumer credit-based culture; we are living beyond our means "eco"nomically—both financially and environmentally.

What we have lost in becoming consumers is the very sustainable art of homemaking. Mr. Berry points out that, in contrast to consumers, homemakers or *householders* (and I don't mean to suggest that only women can be homemakers here) are in some way producers as well as users, providing some of their own needs out of their own resources, skills, and imagination. While householders do buy things, there's a better balance of contributing as well as taking. In learning the domestic skills of cooking, gardening, sewing, building, home remedies, and personal health care, householders become more able, valuable, self-responsible human beings providing the basic necessities of life, with something to give to others and the earth. Instead of being dependent on consuming, householders take pleasure in creating. Households can be places to grow and prepare food, create energy, work, socialize, learn, heal, amuse ourselves, our families, and friends. These activities can be more meaningful and satisfying than working away from home all day in order to be able to indulge in consumer luxuries like the latest fashions and new espresso machines.

Hand-in-hand with householding is the concept of sustenance. *Sustenance* is a means of sustaining life, nourishment; it is that which sustains us. Just as we need to learn how we need to behave to sustain the earth, we also need to learn how to behave to sustain ourselves. By being consumers—destroying, using up, and spending wastefully—we cannot even begin to hope to sustain ourselves or the earth.

What we need to sustain ourselves is clean air, clean water, fertile land, fresh wholesome food simply prepared, practical and attractive clothing, shelter that is appropriate to where we live and what we do at home, meaningful and profitable work, creative expression, loving relationships, participation in community, intellectual stimulation, spiritual growth, and probably a few other things I haven't thought of yet. Our needs for sustenance are basic and simple. But in our consumer culture we sacrifice our sustenance for a fantasy of material fulfillment. Whatever it is we hope to gain by eating packaged foods, wearing the latest fashion, and buying electronic gadgets cannot satisfy the emptiness we have inside when we give up the purity of our air and water, our forests, and biological diversity in exchange. It's having real sustenance in our lives that makes us feel fulfilled and brings us happiness. Consumerism makes us pursue more and more luxuries to fill the emptiness we feel when we lack our basic necessities.

I'm not suggesting that we give up shopping entirely, but rather that we take a different attitude toward how we sustain ourselves—with sustenance instead of consumerism. Sustenance is a hand-sliced loaf of chewy organically grown whole-grain bread, made at home or purchased from a local baker in a recycled paper bag; consumerism is a plastic-wrapped loaf of sliced bleached white bread full of preservatives baked on an assembly line in a factory. Sustenance is saving and investing a percentage of all the money you make; consumerism is using credit to spend more than you make. Sustenance is fixing a special meal at home for your friends; consumerism is meeting friends at a trendy café. Sustenance is exploring every part of your local area—getting to know the people, the wisdom, the arts, the flora and fauna, the weather patterns; consumerism is a first-class trip to Europe. Sustenance is playing the violin in a quartet with your neighbors or being in the audience at a community performance; consumerism is buying CDs. Sustenance is working at something you enjoy for the amount of money you need, plus a little extra; consumerism is doing whatever it takes to get as much money as you can. Again, I'm not suggesting that one shouldn't take a trip to Europe or eat in a restaurant, but rather that focusing on providing our sustenance will lead us to make choices that sustain ourselves and the earth in ways that making choices based on consumerism does not.

Sustaining a home and creating sustenance includes having a living relationship with the land you inhabit and the others who live in your community—the intent being to be responsible for and take care of yourself, your family, friends, and neighbors and the piece of earth you own or share. Sustaining ourselves means sustaining our health, our families, our relationships, our communities, our land, our money, our culture, and everything else that makes life move forward. Material possessions are prescribed by fundamental human necessities and responsibilities to each other and the earth, rather than advertising and cultural conditioning. Living in this sustainable way builds strong local economies that add up to strong national economies, and eventually build into a strong world economy.

To live sustainably requires developing virtues in ourselves that are uncommon in our society: self-restraint, thrift, frugality, nurturing, prudence, wisdom, responsibility, an appreciation of quality over quantity, cooperative relationships, creativity, commitment, and love. Not an impossible task, but one that requires fortitude, thinking for oneself, and responding appropriately, instead of following the dictates of consumer culture.

I don't expect that everyone will suddenly switch from being a consumer to a sustainable householder—that's just not practical. We are all enmeshed in consumer culture, and to live sustainably we will need to create new choices and new systems that support those choices. Meanwhile, though, there are many things we can do to start making the shift away from being rampant consumers to providing for our needs in a thoughtful, responsible way.

First, evaluate what your real needs are. To live sustainably requires a reduction in our use of resources and creation of waste, but not to the extreme. Nature is abundantly productive and part of our role as a species is to be nourished by her gifts. But there is a big difference between taking what we need, using it efficiently, and disposing of it wisely, and taking as much as our available cash and credit allow us to buy, using it wastefully, and tossing it in a landfill. We all have basic needs, both for our physical well-being and human development and our challenge is to fulfill those needs in a way that supports the earth in its long-term ability to provide for us.

Then, if you can make it yourself, make it yourself. Wendell Berry says, "find the shortest, simplest way between the earth, the hands and the mouth." Many products we use daily are *consumables*—those products we use up completely—that can easily be made at home. My husband and I are now preparing most of our food at home from fresh organically grown ingredients instead of buying packaged food or eating out, and we've planted vegetables and strawberries. We also use simple substances for cleaning, such as vinegar and baking soda, and control pests by homemade means. When we're ill, our first choice is a home remedy. Even though we are purchasing many of the raw ingredients we use, we have more control over what goes into the products we make and we've found that the trade-off of spending time is well worth the money we save. Why pay a machine to do something we can do ourselves, and often do better?

If you can't make it yourself, try to buy a local product from a local business. This may be easier said than done, but I've found that if you make a decision to do this, you begin to find local products and services that are not immediately obvious. Most of the food we don't grow we buy through a community supported agriculture program from an organic farmer just up the hill from our house. And, as we ask around, we've found there are many local craftspeople who are making attractive useful things. In our area, there are few outlets for cottage industries to sell their goods, so we're also working on ways in which craftspeople and backyard market gardeners can sell their products to people in the community and how we can share our community skills instead of driving thirty-five minutes to the closest big town. We've also become more aware of which businesses are owned by local residents and which are big corporations, and try to buy from our local small businesses whenever we can.

But unless, as a society, we dramatically change our life-styles, there will always be a need for commerce, manufacturing, and trade. There is evidence that business is moving in the direction of becoming more sustainable, and so when we do make a purchase, it is important that we make the most sustainable choices that are available to us.

One criticism of more sustainable products is that they often are more expensive. This is true is some cases, but not in others. The most sustainable products are often comparable in price to the highest-quality, top-of-the-line unsustainable products, but many are competitively priced or only slightly more expensive. I've found for myself that even though I often spend more per product, I actually spend less money overall than I did before because I'm buying fewer things. After eliminating the products I don't need and making some items myself, I make it a priority to support those businesses that are making positive environmental change. Even though individual products may cost more, a sustainable life-style overall costs less.

Before you become immersed in the product information in this book, there is one final thought I want to leave you with: Assess products for their sustainability, but don't be too picky. There are few perfect products, but many that are more sustainable than others. Look at the environmental attributes of a product as a glass half-full, rather than the problems of a product as a glass half-empty. Acknowledge companies that are working toward sustainability with your purchases, so they can continue to make improvements. Instead of looking for reasons to reject products, look for reasons they are better. The nature of our free-market system is to respond to consumer demand—let's lead the market in the direction we want it to go.

HOW TO USE THIS BOOK

This consumer guidebook has two parts. It is designed to explain the process of product evaluation and give guidelines and guidance on choosing specific products.

Part One contains background information about applying the concept of sustainability to all phases of a product life cycle. It explains how products are labeled to indicate environmental attributes, common buzzwords, and environmental issues that can be improved by our product choices.

Part Two gives guidelines for finding and choosing specific products we commonly use at home, listed in alphabetical order. If you're primarily interested in choosing the most sustainable shampoo, for example, you can just look it up under SHAMPOO AND CONDITIONER in Part Two.

Products indicated in the text of Part One with boldface small caps are included in Part Two.

Part One

CHAPTER 1

Sustainability: The New Benchmark for Evaluating Everyday Products

In response to both consumer demand and environmental need, over the last few years there has emerged a genuine interest and inquiry into determining the real environmental impact—both positive and negative—of the products we use every day as we go about our lives. While there are many points of view regarding how one might make this environmental assessment, ultimately it's a matter of sustainability.

In the realm of everyday products, sustainability is a solid benchmark amid a sea of often contradictory claims. One manufacturer tells us that it's good for the environment to use disposable diapers because you save water by not having to wash them, while another tells us cloth diapers reduce landfill waste. The plastics industry advertises that plastic saves energy in shipping because it's lighter than glass, but the glass bottle is both made from recycled material and is recyclable after use. So which really is better? Sustainability gives us the answer.

The concept of sustainability starts with the acknowledgment that we humans are a part of the natural world. It is rooted in looking to the inherent workings of nature as a model, with the idea that the natural systems of the world work in balance to perpetuate life, and by working in harmony with those natural systems, we can sustain our own lives. Put simply, sustainability—as it relates to the environment—is the ability to keep our ecosystem going over time, taking from the earth and giving back to the earth in balance.

It's a law of nature that living organisms must have an exchange of raw materials and wastes with the surrounding environment in order to survive. The human population and economy depend upon constant flows of air, water, food, raw materials, and fossil fuels from the earth (called "sources" in ecology jargon), and we constantly emit wastes and pollution back to the earth (called "sinks"). There are limits to the rates at which the human population can use materials and energy and there are limits to the rates at which wastes can be emitted without harm to the people, the economy, or the earth's processes of absorption, regeneration, and regulation. Sustainability is a balance of taking and giving—not all taking or all giving, but both appropriately in a way that sustains the continuation.

Each resource used by humans—food, water, wood, iron, oil, and many, many others—is limited by both its sources and its sinks. The exact nature of these limits is complex, because sources and sinks are all part of the dynamic, interlinked, single system of the earth. Some limits are much more stringent than others. There are short-term limits (such as the number of apples on a tree) and long-term limits (such as the amount of oil underground). Sources and sinks at times interact, and the same natural feature of the earth may serve as both a source and sink at the same time. A plot of soil, for example, may be a source for food crops and a sink for acid rain caused by air pollution.

There's nothing inherently wrong with taking from and giving to the environment, but we must do so in a sustainable way that maintains the ability of our surrounding environment to continue to give and take. Right now we are on the brink of going beyond our limits; some say we have already stepped over the line.

We simply cannot continue to extract resources from the earth and emit pollution at the rates we have done in the past. Without significant reductions in the natural resources we consume and pollutants we produce, in the coming *decades* (within our and our children's lifetimes) there will be an uncontrolled decline in food, energy, and all kinds of consumer goods that we take for granted today. Yet a change to sustainability is still technically and economically possible. We still have a choice. We still have time to reverse our direction before it is too late.

We can achieve sustainability when we act in such a way as to take only the amount of resources that the earth can regenerate, and dispose of only the amount of pollution and waste that the earth can assimilate. There is only one question, now, to ask, only one benchmark, one guideline: Does it contribute to sustainability or does it lead away from sustainability?

SUSTAINABLE PRODUCTS

A product is sustainable if it is made, used, and disposed of in such a way that it could continue to be made, used, and disposed of again and again indefinitely. Sustainability defines two basic product criteria:

1. For *sources*, a product must utilize natural resources in a way that allows those resources to be available from generation to generation.
2. For *sinks*, the waste from a product must stay within the manufacturing loop or assimilate into the natural ecosystem and not build up or cause pollution.

This sounds simple enough, but in reality assessment for sustainability becomes very complex because there are so many factors to consider. Consequently, right now there are few completely sustainable products on the market, and many with some degree of sustainability. The majority of products available still, however, is almost completely unsustainable and there lies our problem: How can we sustain life on earth if the majority of what we purchase and how we live contributes to its destruction?

Our task as shoppers is to identify which products available to us are most sustainable and use them, so that more and more sustainable products are introduced into the marketplace. Identifying products with sustainable attributes is fairly easy, once you know what to look for. Before I explain the different aspects of choosing sustainable products in detail, I want to give a brief overview of the many factors that should be considered when evaluating the sustainability of a product.

CHOOSING PRODUCTS FOR SUSTAINABILITY

Here is the list of questions I go through when I am making product choices in the marketplace. They give me general clues as to the product's environmental sustainability and help me decide if the product is an appropriate choice for me. While a scientific product analysis is much more complex than this (see Chapter 3), at this time, this list of questions is one we can use while shopping to help us make a choice about a product based on the information immediately available on the label.

1. **Do I really need the product?** I always ask myself this first, because every product I don't buy saves resources and eliminates waste. When

I took a good look at what I was buying, I found that I was buying a lot of things I really didn't need, even though they were "green." Now I buy only (well, almost only) things that satisfy a functional and/or aesthetic need and are not superfluous clutter. When appropriate and possible, I buy a secondhand product over a new one (such as antique furniture or used books).

2. **Is the product safe to use?** Because I value my health and am very aware of the dangers of toxics in products, once I decide I am interested in a product, this is my first question. If the answer is no, it goes back on the shelf. Because there are so many nontoxic alternatives available, this is one area in which I personally will not make a compromise, unless there is absolutely no other choice.

3. **Is the product practical, durable, well made, of good quality, with a timeless design?** Considering the environmental cost to make every product, I want a product to last. So I avoid the latest trendy fashions or designs and cheap imitations and buy better-quality goods. I could buy cheap things at discount stores, but I've found that taking the time and care to choose well-made goods more than pays for itself in durability.

4. **Is the product made from renewable or recycled materials taken in a sustainable way?** Are the ingredients/materials listed? What are the ingredients/materials used that are not listed? Any warning labels or environmental claims? In looking at the life cycle of the product itself, I first ask myself if the raw materials are renewable (plant, animal, earth) or nonrenewable (metals and petrochemicals). If the product is made from paper, glass, metal, or rubber, I look for recycled content. Then I also particularly look for labels that indicate whether the materials used are organically grown, sustainably harvested, or other explanations that describe sustainability.

5. **Is there any information about the manufacturing practices that tells of environmental improvements?** This is the area where there is the greatest environmental impact and we as consumers have the least information. You probably won't have the exact data needed to evaluate this step in the product life cycle, but you can make assumptions based on general knowledge (such as that recycled paper uses less water and energy in the manufacturing process than virgin paper).

6. **How will I dispose of the product, and what environmental impact will that have?** I look for products that are biodegradable, and if I can't put it in my compost pile or let it safely run down the drain, I want it to be recyclable in the area where I live.

7. **What kind of package does the product have?** I prefer products that can be purchased in bulk or that have no packaging and can be slipped into the fold-up cotton bag I carry in my purse. I try to avoid plastic packaging.

8. **How far has the product been shipped to reach the retail outlet?** I try to find products that are made as close to my home as possible. Given the choice between two products with equal environmental merit, I'll choose the locally made one to cut down on resource use and pollution from transportation, and to support my local economy.

9. **Is the product a good value for the money?** While my personal feelings are that I want to buy the best environmental option I can find, there are some products on the market that are just too expensive for what you get. While I believe that the protection of the environment is worth any price, there are some products that are a waste of money, or don't warrant the added cost.

10. **Is there some environmental, health, personal, or economic benefit that outweighs the product's environmental cost?** This is really the final question, because there may be other factors that justify a compromise. An environmental benefit might be that the product saves energy or water, or produces solar power. A health benefit could be that the product filters contaminated water, or purifies indoor air. You may need to purchase a book printed on virgin paper with toxic ink that is necessary for your personal growth or helps you plant an organic garden, or you may not have the time to run across town to buy an energy-efficient light bulb at the moment your old light bulb burns out. Economical reality may be that you cannot afford to spend eighty dollars a gallon on the most natural wall paint, or even a little extra for organically grown food.

EVALUATING PRODUCTS IN THE REAL WORLD

You can ask yourself the questions listed above for virtually any product you are considering. As an example, using the tables that follow, let's look at a product choice we make every week. "Paper or plastic?" when asked at the checkout counter takes on new meaning in the context of sustainability. Let's assume that the paper bag is certified by Scientific Certification Systems (see Chapter 2) to contain 40 percent post-consumer recycled paper, the plastic bag has no labeling, and we also have the choice of a cotton bag made from unbleached cotton (not organically grown) with a store logo printed on it.

PLASTIC	PAPER	COTTON	QUESTION
Yes	Yes	Yes	**1. Do I really need or want the product?** Yes, I need something to carry my groceries home in, so I have a legitimate need for any of the three options.
Yes	Yes	Yes	**2. Is the product safe for me to use?** There are no negative health effects I am aware of for any of the options. Plastic bags should be kept out of reach of children, but I don't have a child.
		Best	**3. Is the product practical and durable, well made, of good quality, with a timeless design?** The paper bag could be reused probably twenty or thirty times if I keep bringing it back to the store; the cotton bag could be reused indefinitely until it wears out (let's say five years or more); and the plastic bag could be reused indefinitely, but it's not as durable, so won't last as long. The most practical and durable choice here is the cotton bag.
No	Better	Better	**4. Is the product made from renewable or recycled materials taken in a sustainable way? Are the ingredients/materials listed? What are the ingredients/materials used that are not listed? Any warning labels or environmental claims?** The plastic bag has no ingredients listed or environmental claims, though it is obviously plastic and I know that plastic is made from crude oil, a nonrenewable resource in limited supply—no long-term potential for sustainability. The paper bag has a large percentage of recycled post-consumer content, which is good, but more than half the paper pulp comes from cutting down trees (the forestry practices are not known, so I assume they were clear-cut). The cotton was grown with pesticides but no dyes were used. Both the paper and cotton bags have printing on them. Here I would say it's a toss-up between the cotton and the paper. Both are renewable, but both use manufacturing practices that are not particularly sustainable.

PLAS-TIC	PAPER	COT-TON	QUESTION
—	—	—	**5. Is there any information about the manufacturing practices that tells of environmental improvements?** The only product that gives any indication of manufacturing practices is the paper bag with the recycled content. If we were comparing paper bags, this one would come out ahead over a bag with no recycled content because of the savings of water and energy. Since comparable information for the other products isn't available, we can't really answer this question.
No	Better	Better	**6. How will I dispose of the product, and what environmental impact will that have?** The plastic bag will have to go in the trash and go to a landfill. Some plastics can be recycled in some places, but where I live, there are no plastics recycling facilities. The paper bag can be recycled. The cotton bag can also be recycled, or ripped apart and placed in our compost pile.
—	—	—	**7. What kind of package does the product have?** None of the bags comes in a package.
—	—	—	**8. How far has the product been shipped to reach the retail outlet?** None of the bags has information on where they were manufactured, so this question cannot be answered.
		Best	**9. Is the product a good value for the money?** Plastic and paper bags are "free." Most markets in my area give a five-cent credit for every bag you supply for your groceries. Say the cotton bag cost $5.00 (there are many that are even less). You would have to use the bag 100 times to have it pay for itself. If you went to the market once a week, after two years the bag would pay for itself. We have about four bags of groceries a week. After the two-year payoff, we would save about $10 a year by bringing our own bags. That's not a lot, but over ten years it's $100. So I would say in the long run, the cotton bag is the best value.

PLAS-TIC	PAPER	COT-TON	QUESTION
—	—	—	**10. Is there some environmental, health, personal, or economic benefit that outweighs the product's environmental cost?** In this case, no. I'll make this choice based on the environmental benefits.
2	4	6	**And the winner is . . .** Based on the factors we were able to compare, the cotton bag came out on top, with the recycled paper bag a close second. If the cotton bag were made of organically grown cotton or the paper bag had no recycled content, the difference would have been greater.

Using this checklist can help when choosing among products. It is very simplistic, considering the vast number of details that could be evaluated for any given product. However, it is useful, and appropriate for the amount of information about products that is currently available from manufacturers. Making some choices will be easy, like choosing between a toxic drain cleaner with lye that will severely burn your skin if you spill a drop, or one that's nontoxic. Other choices will be more difficult because of the complex web of our personal needs versus environmental needs.

A friend and I had a debate one day about the environmental merits of using an electric food processor that uses nonrenewable energy and is made of plastic. I maintained that food could be processed with a human-powered knife made from recycled steel, a grater, and other manual tools. He argued that having a food processor enabled him to prepare food quickly for his family from fresh ingredients, which greatly reduced the amount of packaging he would otherwise use in buying convenience foods, and energy used to ship these products, as he had a large garden and also purchased many fresh ingredients from nearby roadside stands. Because he could chop, grate, puree, slice, and otherwise process raw foods quickly, his family ate more healthfully.

The real issue here is not to compare a knife and food processor but rather how we need to prepare our food in order to eat it. The most sustainable choice would be the one that requires the fewest resources and creates the least amount of waste and still produces the desired result of getting the food from your hand to your stomach: simply eating the food whole. However healthy and sustainable this might be, few of us eat only whole, raw foods, though we occasionally enjoy them. Instead of munching on a carrot, we're more likely to grate it into a salad, slice it into crudités and dip it in a dressing, chop it up and put it into soup, or cut it up and steam it with butter or a glaze. These preparations require both energy and various tools, depending on the complexity and refinement of your cuisine. It's difficult to make a vegetable puree soup, for example, without a food processor, but do you need to puree your soup? I just make mine chunky and cut the vegetables with a knife (my friend still uses his food processor).

All product choices will have such trade-offs. The most important thing is that we begin to move in the direction of environmental sustainability, and support those products that have made real environmental improvements to move the marketplace in that direction. Just take this a product at a time, and soon looking at the environmental attributes will become second nature.

Without a sustainable base of resources from which we and future generations can continuously draw, we cannot survive. Everything we have, indeed, our very lives, begin and end with the earth. There is no greater priority than to sustain that which sustains us.

Conservationist Aldo Leopold said, "A thing is right when it tends to preserve the integrity, stability, and beauty of the biotic community. It is wrong when it tends otherwise." Let this sentiment guide our product choices.

CHAPTER 2

Reading Labels

When the first few products came on the market claiming to be environmentally friendly there were no regulations or guidelines for correctly labeling these products. Manufacturers could and did state anything they wanted to on their labels in order to make it appear their products were better for the environment.

Quickly, however, it became apparent to both consumers and regulators that such labeling was not only misleading, but there was so much confusion that manufacturers and marketers didn't know what to say about their products and consumers didn't know what to believe or buy. Some companies continued deceptive labeling practices, while others—wanting to avoid labeling problems—decided not to make claims at all, even though their products did have environmental benefits. Without adequate and correct labeling we cannot know the facts about a product or make wise purchasing decisions.

Good labeling is vital for us to be able to choose products that are more sustainable—I'm finding more and more that as new products with environmental attributes come on the market, their labeling is clear and informative.

A GOOD ENVIRONMENTAL LABEL

Even though there is little consistency in environmental product labeling, and there is a lot of room for improvement, product labels can give us both clues and, in some cases, detailed information to help us identify products that have true environmental attributes and reject those that are trying to take advantage of our environmental concern. Here's how to identify a good environmental label.

ENVIRONMENTAL CLAIMS SHOULD BE SPECIFIC, NOT VAGUE OR INCOMPLETE. Avoid products that only say "environmentally friendly," "environmentally safe," "good for the planet" or have a similar message without giving any information about why the product actually is good or better for the environment. These terms are inaccurate, too vague to be meaningful, and impossible to substantiate. Also remember that just because a brand name uses words such as *eco, green, earth,* or *planet* that does not mean it necessarily has legitimate environmental benefits. Look for truthful, specific claims that specify the precise environmental attribute of a product, such as "made from 100% unbleached post-consumer recycled paper (recycled newspapers)." Clear, easy-to-understand language should be used and it should be written in a legible type size.

A COMPLETE LIST OF INGREDIENTS CONTAINED IN THE PRODUCT SHOULD BE GIVEN, ALONG WITH ANY RELEVANT MANUFACTURING AND DISPOSAL INFORMATION. Companies should make information about the composition and the environmental effects of their products available. In addition to the product label, many companies will provide, upon request, literature on the environmental benefits of their products.

ENVIRONMENTAL CLAIMS SHOULD BE SUPPORTED BY COMPETENT AND RELIABLE SCIENTIFIC EVIDENCE. The Federal Trade Commission Policy Statement on Advertising Substantiation states that advertisers must have a reasonable basis for making express or implied claims that contain objective assertions about the product or service. Claims must be supported, whether by tests, analysis, research, a specific study, a set of studies, or a consensus of opinion.

A CLEAR DISTINCTION SHOULD BE MADE BETWEEN THE ENVIRONMENTAL ATTRIBUTES OF A *PRODUCT* AND THE ATTRIBUTES OF THE *PACKAGE*. Check the label carefully to determine if the environmental attribute or benefit refers to the product, the product's packaging, or to a portion or component of the product or packaging (generally if the attribute applies to all but minor, incidental components, the claim is not so qualified).

CLAIMS SHOULD NOT OVERSTATE THE ENVIRONMENTAL ATTRIBUTE OR BENEFITS, EITHER DIRECTLY OR BY IMPLICATION. Use common sense to determine if a negligible environmental benefit is presented as being significant.

COMPARATIVE STATEMENTS SHOULD BE PRESENTED IN A MANNER THAT MAKES THE BASIS FOR THE COMPARISON SUFFICIENTLY CLEAR TO AVOID

DECEPTION. Comparisons should include exact percentages in reduction of weight or volume and state clearly what it is being compared to. In addition, the advertiser should be able to substantiate the comparison.

CERTIFICATION OF CLAIMS SHOULD BE SPECIFIC. Look for certifications by reputable organizations—such as Scientific Certification Systems, Green Seal, or local certifying agencies for specific types of products—that give meaningful indicators that the product meets certain standards.

INDEPENDENT CERTIFICATION PROGRAMS

As more and more products with environmental claims come on the market, with them comes the development of organizations to independently certify that their claims are true. As of this writing, environmental product certification programs were operating or under way in eleven foreign countries, the oldest and most famous being the Blue Angel program, which originated in what was once West Germany.

The goal of the foreign certification programs is to motivate manufacturers to voluntarily produce products that are environmentally benign, and to help consumers who are seeking such products to identify them by a clear, readily recognizable seal of approval.

These programs set high product standards, invite manufacturers to voluntarily apply to meet the standards, identify products that do meet the standards, and grant the manufacturers of qualifying products the right to use the program's identifying label or seal on the products, for a fee. Although foreign certification programs are generally established under the authority of the government, nongovernmental groups are involved in the actual design, substance, and operation of the program. Participants often include manufacturers, retailers, environmentalists, trade unions, and consumers.

In the United States, we do not have such a government program—instead we have two independent certifying organizations for general environmental products in the private sector (Scientific Certification Systems and Green Seal), as well as a number of smaller organizations that certify only a single type of product, such as organically grown food or sustainably harvested wood (see Chapters 5 and 7). Each has its own systems for evaluating and certifying environmental claims on products and packaging, as well as standards which must be met to display their respective seals. These indicate to consumers that the product or package is a good environmental choice. The value of a certification, to both manufacturers and consumers, is to lend credibility to a claim by having it verified by an independent third party.

Under both the Scientific Certification Systems and Green Seal programs, the manufacturer enters into a contract with the certifier and pays a fee for the certification services. The evaluation process may require product testing and a factory inspection and audit. Depending on the results, certification is either issued or denied. If certification is granted, the manufacturer has the right to refer to the product certification program and display the program seal on products, packaging, and advertising. Both programs review the product claim on a quarterly basis for an additional fee. In addition to environmental concerns, both also have high performance requirements (cleaning products, for example, really have to clean).

Both Scientific Certification Systems and Green Seal have made a major contribution to helping consumers, as the research necessary to create their standards has brought a new level of inquiry and understanding that benefits all of us, whether or not their seal of approval appears on a product. In the beginning Scientific Certification Systems and Green Seal were fairly competitive; however, over time their programs have developed to use altogether different approaches to certification and to certify different types of products. We can rely on both programs to certify products that are above average in their environmental integrity.

Don't assume, however, that just because a product has a certification it is superior to other products of the same type. A certification simply means that a manufacturer has been willing to pay a fee to be able to say that a third party has verified that the product meets certain criteria. As consumers, it is as important for us to know the criteria as it is to see the certification so we can compare the certified products to those that have chosen—for whatever reason—not to be certified. A certification also doesn't mean that a product is completely safe for the environment—it means only that a particular claim has been verified or a certain standard met. And you need to use your own common sense—a paper or plastic grocery bag certified to have recycled content is not necessarily better for the environment, all things considered, than your own uncertified reusable cotton bag.

Scientific Certification Systems

Scientific Certification Systems (SCS) states as their mission a commitment to "developing programs that motivate private industry to work toward an environmentally sustainable future." To accomplish this, SCS has "established scientific procedures for conducting independent, unbiased evaluations of products and product claims, and for recognizing products achieving exceptional environmental performance goals."

The organization was established in 1984 to develop certification procedures for food safety claims (see Chapter 5), and began certifying environmental claims in late 1989 using the name Green Cross (the name Green Cross is no longer

used, but the green cross still appears on their logo). To maintain total independence and objectivity, the company was funded entirely by its founders, and does not have ownership interest of any kind in any of the products it certifies.

SCS awards certification on two levels.

First, they developed a program to verify environmental claims relating to specific product attributes. SCS has developed standards to certify claims of recycled content, sustainable forest management, reduced volatile organic chemical air emissions, and biodegradability (see Chapters 6, 7, 10, and 11).

You may have already noticed a SCS certification on your paper grocery bags, or on a store shelf talker (some brand name certified products are listed in relevant product listings in Part Two). They also certify packaging such as plastic bottles with recycled content and recycled paper egg cartons, and raw materials used to make products.

Upon certification, recipients are awarded a certificate of achievement and are entitled to print an informative certification message, accompanied by the Scientific Certification Systems logo, directly on their products and packaging. The informative statement specifically describes each environmental attribute that has been certified. They emphasize that they are not certifying the product as being better for the environment, but rather they are verifying the truth of the environmental claim. Consumers must decide for themselves whether or not a particular attribute (such as recycled content) is actually beneficial for the environment.

SCS also offers a deeper level of certification—their Environmental Report Card—based on a "life cycle" examination of the full range of environmental effects associated with the manufacture, use, and disposal of a product and all of its associated packaging and shipping (see Chapter 3).

In addition to working with manufacturers, SCS has ties with retailers such as Home Depot, to review the environmental claims made by some of their suppliers. They also offer other technical support services, such as working with advertising staff to review the use of environmental claims in their advertisements, and assisting in the development of corporate environmental policy.

An independent Scientific Advisory Board comprised of individuals from business, academia, and the environmental community has the authority to review all SCS programs. In addition, all SCS labeling programs are subject to public and peer review. The Good Housekeeping Institute works with SCS in the development of efficacy guidelines.

Green Seal

Green Seal was established in June 1990 as an independent, nonprofit environmental certification and education organization. Its founder and former president is Denis Hayes, executive director of the original Earth Day in 1970 and international chairman of the phenomenally successful Earth Day 1990. Hayes has since gone on to fight other environmental battles, but Green Seal continues.

The board of directors for Green Seal includes, among others, representatives from business, environmental, consumer, and other public interest organizations. Manufacturers are represented on Green Seal Advisory Committees. The organization's Environmental Standards Council is composed of scientists and other experts and acts as a certification appeals board. To protect the integrity of the Green Seal certification mark, board members and staff are required to sign a "Code of Conduct" that prohibits financial conflicts of interest.

Underwriters Laboratories, Inc., conducts the majority of Green Seal's testing and factory inspections and is responsible for follow-up inspections to monitor compliance with Green Seal's standards.

The Green Seal certification mark is awarded to products that meet strict environmental standards set on a category-by-category basis. Green Seal conducts an "environmental impact evaluation," which identifies the significant environmental impacts of a product and its packaging from the manufacturing stage through consumer use to recycling and disposal. The organization then develops standards designed to reduce the product's major adverse impacts.

Some of the impacts that Green Seal may consider for evaluation in establishing a standard are toxic chemical pollution, energy consumption, depletion and pollution of water resources, harm to wildlife and natural areas, waste of natural resources, destruction of the earth's atmosphere, and global warming. All the standards have packaging requirements to minimize resource use and toxic materials in inks.

Green Seal differs from Scientific Certification Systems in their approach because their seal indicates compliance with a multifaceted set of standards, rather than a single attribute certification or a complete life cycle analysis. Nei-

ther approach is more correct than the other, but it's important to know that they are different.

When a proposed standard is ready, Green Seal circulates it to manufacturers, trade associations, environmental and consumer groups, government officials, and the general public for comment. Following a review of the comments, Green Seal issues a final standard for evaluating applicant products.

Green Seal has issued final standards for facial and bathroom tissues, paper towels and napkins, re-refined engine oil, water-efficient fixtures, compact fluorescent lamps, printing and writing papers, household cleaners, paints, clothes washers, clothes dryers, and dishwashers (some brand-name certified products are listed in relevant product listings in Part Two), and will review these standards every three years. In process are standards for coated printing paper, energy-efficient windows and window films, newsprint, reusable utility bags, sealants and caulking compounds, adhesives, recycled toner cartridges, refrigerator-freezers, cooktops and ovens, and room air conditioners.

A manufacturer that is awarded the Green Seal may use it on a product and its packaging and in product-specific advertising and promotional materials. A description of the basis for certification must appear with the certification mark.

As well-researched as the Green Seal standards are, it is important to remember that they must set their standards at a level where there are products that can be certified. Therefore, the standards do not reflect the most sustainable possibility, but rather a step in a more environmentally beneficial direction. Many of the standards (described throughout the book) focus on just a few aspects of the product. Household cleaners, for example, are evaluated for toxicity and biodegradability, but no mention is made as to whether the resources used are renewable or nonrenewable. It is likely that the standards will be revised many times in the years to come to reflect both new environmental knowledge and increased ability of manufacturers to produce more sustainable products.

MARKETING PLOYS TO WATCH OUT FOR

Because there are so many products on the market now making so many environmental claims, it's important that we as consumers have the background knowledge and information to be able to choose those products with legitimate environmental benefits rather than those that are simply using our concern for the environment as a marketing ploy.

The most important thing to remember is to look for products where the product itself has environmental benefits. Many manufacturers have, in fact, changed their products to be better for the earth. Recycled-paper toilet paper, tissues, and paper towels are now on supermarket shelves, and many new natural cleaning products have been formulated. Nontoxic pesticides are carried in most nurseries, and hardware stores have energy-efficient light bulbs. These products were not commonly available only a few years ago (see Part Two for product descriptions).

But there are also products with somewhat less environmental integrity that are being target marketed to environmentally concerned consumers. Since the concept of shopping for the environment first became popular in the late 1980s, manufacturers and marketers have made their products seem more "green" in a number of different ways, even if the product itself has no real environmental benefit and may even cause significant environmental harm. Here are a few tactics to be aware of when looking for sustainable products.

DECEPTIVE OR UNSUBSTANTIATED CLAIMS. As widespread consumer interest in the environment became apparent, manufacturers wanting to cash in on this market used any reason to claim their products were green. One manufacturer of polystyrene hot-cold cups (which are made from unsustainable, nonrenewable crude oil, and are not biodegradable when disposed of) claimed its product was better for the environment because it "saved trees." While it is true that it saved trees, it is not a legitimate claim to justify using a product with its own environmental problems. One manufacturer that claimed its plastic trash bags were biodegradable was later taken to court by the Federal Trade Commission for making a claim it couldn't substantiate. Because of government intervention, such labeling rarely occurs today, but it's still important to watch out for bogus claims.

GIVING MONEY TO AN ENVIRONMENTAL ORGANIZATION. While charitable giving is a worthwhile thing to do—especially if it helps support environmental protection or restoration—it doesn't make the product itself better for the environment. In fact, it is often products that are the most environmentally destructive that attempt to lure consumers with this kind of information. Don't get me wrong—if the only choices on your supermarket shelf are two equally toxic insecticide sprays, I'd rather have you buy the one that makes a donation to an environmental organization—but the donation hardly makes amends for the environmental pollution and health hazards created by the manufacture, use, and disposal of the pesticide. A better choice would be to look for one of the safer, natural pest controls now commonly sold in nurseries and hardware stores.

CHANGE IN PACKAGING. One of the most pressing environmental problems in our country today is solid waste disposal. A very practical solution to the problem of overflowing landfills is to simply reduce the amount of trash we create. Over the past couple of years manufacturers have made great progress in reducing the amount of packaging surrounding a product, changing the product itself so that less packaging can be used (such as the reformulation of super-powered laundry detergents that fit the same amount of cleaning power in a smaller box), and making packages out of recycled paper, plastic, metals, and glass that can again be recycled or refilled instead of going to a landfill. But a sustainable package does not make a sustainable product. If you are choosing between, say, two boxes of cereal, I would prefer you choose the one in the recycled paperboard box over the one in the plastic bag. However, an improvement in the package does not automatically take care of the environmental problems caused by the spraying of pesticides on the cereal's ingredients or the loss of topsoil caused by agribusiness farming methods. Manufacturers deserve credit for improving their packaging, but their products need to be better for the environment too.

FEDERAL AND STATE REGULATIONS

The federal government has few legally binding guidelines or regulations for environmental labeling. Currently the use of environmental marketing claims in advertising and labeling is regulated primarily by the Federal Trade Commission and the Environmental Protection Agency.

The Federal Trade Commission (FTC) was created by the Federal Trade Commission Act of 1914, which empowered the agency to prevent unfair or deceptive activity, including advertising, and to enforce federal laws on price-fixing and monopolies. In addition, the agency could issue trade regulation rules based on unfairness or deception. In 1980 Congress rewrote the Federal Trade Commission Act to allow the FTC to issue trade rules based on deception, but not on unfairness.

The FTC's Consumer Protection Bureau is the federal agency charged with the enforcement of federal "truth in advertising" laws. Even before the Task Force of Attorneys General was formed, the FTC announced it was investigating products making environmental claims.

The FTC has neither the expertise to make environmental evaluations nor the statutory authority to establish environmental policy. It does, however, have the authority to ensure that environmental marketing claims are not deceptive or unfair, and that such claims are adequately substantiated. Under statutes prohibiting unfair competition and deceptive, unsubstantiated, and unfair advertising, it can take legal action if environmental claims are not clear and accurate.

Over the past few years the FTC has conducted more than two dozen investigations into the use of environmental marketing claims: a plastic-coated paper milk carton claimed to be "biodegradable"; a spray cement containing an ozone-depleting chemical claimed to have an "ecologically safe" propellant; plastic trash bags claimed to be "degradable," "photodegradable," and "biodegradable."

In July 1992 the FTC issued voluntary guidelines for environmental advertising and marketing practices. Prepared with the aid and approval of the EPA and the U.S. Office of Consumer Affairs as part of a Federal Interagency Task Force on Environmental Labeling, the FTC guidelines offer industries an interpretation of how existing FTC policies on deception, advertising substantiation, and unfairness are likely to be applied specifically to environmental marketing claims. The goal of these guidelines is to protect consumers, to bolster consumer confidence in environmental claims, and to reduce manufacturers' uncertainty about which claims might lead to FTC law enforcement actions. The guidelines are strictly voluntary and do not carry the weight of law. The FTC will request comments on how the guidelines are working in 1995.

The Environmental Protection Agency (EPA) does not have statutory authority to create binding regulations in the area of environmental labeling. Through the Federal Interagency Task Force on Environmental Labeling, the EPA has worked with the FTC in enforcement actions against product manufacturers concerning certain environmental marketing claims, and also has joined with state governments and with the Coalition of Northeastern Governors to address the regulation of environmental marketing claims. In addition, the EPA has developed federal procurement guidelines for certain products that contain recycled material (see Chapter 6).

The EPA has taken the position that environmental marketing has an important role to play in educating consumers about the potential impact that everyday, individual choices in the marketplace can have on the fate of the planet. Their goal is to encourage environmental marketing. Because they fear that widespread consumer confusion and skepticism about environmental claims could undermine the market and incentive for environmental products, the EPA supports the development of consistent federal guidelines for the use of certain environmental marketing terms, as well as a unified federal enforcement strategy.

In the absence of any federal regulations, individual states have begun to issue regulations for environmental labeling, packaging, and recycling. These take many forms—from mandatory regulations to voluntary guidance—and address issues from truth in advertising to toxics in packaging and organization of recycling programs. Arizona, California, Connecticut, Florida, Illinois, Indiana, Iowa, Maine, Massachusetts, Minnesota, Missouri, New Hampshire, New Jersey, New York, Rhode Island, Tennessee, Texas, Utah, Washington, and Wisconsin all have laws on the books that can be enforced.

Because these regulations are not consistent from state to state, compliance is difficult, time-consuming, and expensive. Some of the regulations simply result in manufacturers removing often legitimate environmental claims from their labels because it is too difficult to comply with the different regulations. Others, such as regulations on packaging reduction or materials bans, are mandatory and must be complied with in order for a manufacturer to sell its products within that state. Industry trade groups want federal regulations to eliminate conflicting state regulations.

In the chapters that follow, I make many references to state regulations. These are only meant to be examples—space does not permit listing every law in every state and new regulations on environmental labeling are sure to develop. Since consumers do need to know regulatory information on a state-by-state basis, I'd like to see a book of applicable regulations be a project of a local consumer or environmental group.

CHAPTER 3

Product Life Cycle

One of the difficulties of determining the sustainability of a product is that there are so many variables to consider. When we look at a product and see that it is biodegradable or made of recycled material, this is only one attribute of the product. In fact, there are many, many variables in the production and use and disposal of a product from the moment the raw materials come from the earth to the time the wastes are disposed of, and each step of the way might sustain or harm the environment. In order to be truly environmentally sound, a product should be sustainable in every way.

The process of looking at the entire life cycle of a product is called, appropriately, life cycle assessment (LCA). The basic idea is to identify and evaluate all the environmental impacts of a set of linked activities that comprise the product from "cradle to grave." This sounds quite logical, but in practice is so difficult that standard methodology is still being debated. LCA is generally a four-stage process: First the perimeters of the LCA system are set, then the current life cycle data is gathered to make an inventory, next an analysis is made of the environmental impacts, and finally improvements are suggested to make the product more environmentally safe.

There are several benefits to doing life cycle assessment, but most important is that it give a multidimensional view of the product instead of looking at only one issue. If, for example, you choose a product solely on the basis of its being nontoxic during use, you might completely miss the fact that it is made from a nonrenewable resource, or that its manufacture creates toxic waste. LCA is like watching a movie of the life of a product, rather than looking at a still picture of only one environmental attribute; you get a truer picture of the product when you look at the whole film instead of just one frame.

But there are also limitations. LCA can be and has been used quite effectively to compare inventories of materials used or pollutants generated within a defined system (such as Brand X uses less energy to manufacture than Brand Z), but there is still much work to be done on the analysis of the environmental impacts (what, for example, are the myriad, interconnected environmental effects of dumping toxic waste into a river?).

When used by industry, LCA also does not generally take into account qualitative differences, such as use of a renewable resource versus a nonrenewable resource, or answer such questions as "Is it better for the environment to put disposable diapers in a landfill, or use nuclear energy and scarce water to wash and rewash cloth diapers?" (When compared this way, both options might look equally bad, but from a sustainability viewpoint the cloth diaper is better—though not perfect—because it is reusable and does not require the materials, energy, and water that go into the manufacture of each and every disposable diaper. Also, disposable diapers nearly always end up in a landfill, but cloth diapers can be washed with safe, renewable energy, and water is abundant in some areas.)

LCA also cannot measure such factors as the amount of managed land needed to provide a resource, or the environmental implications of land management, including loss of species diversity or soil erosion. LCA needs to include subjective as well as objective data—simply measuring the ins and outs doesn't always give the whole picture.

Despite the controversy and disadvantages, LCA is the most useful framework currently available for organizing the many bits and pieces of information about a product, and gives clues as to the questions we should be asking. In the not too distant future I believe all manufacturers will have to have some kind of life cycle information regarding their products available to their customers.

LCA also shows clearly why it is meaningless for any product to make a blanket statement that it is "environmentally friendly" and why it is impossible to have a single seal of approval indicating that a product is good for the environment. There are just too many factors to consider.

At this time there is no reliable, meaningful, generally accepted methodology for conducting complex product life cycle assessments, though the Environmental Protection Agency and other private and public organizations are working toward that goal.

THE LIFE CYCLE OF A PRODUCT

While to document every little detail that goes into the making of a product is incredibly complex, the basic structure of a product life cycle is fairly simple to chart.

There are two simple concepts at work. The first is to measure the inputs and outputs that go into a product system. For example, if you were baking a cake, you would input eggs, flour, milk, sugar, and other ingredients into a bowl, use energy through the electric mixer to process the ingredients and energy in the oven to bake the cake, and output a cake and some dirty dishes. Manufacturing a product is a very similar but more complex process.

The other factor is the different stages a product goes through beginning with the taking of the raw materials from the earth and ending with the disposal of the product after use. Each of these steps has inputs and outputs, and each must be analyzed for its sustainability:

- Raw materials acquisition
- Manufacturing
    - Processing of raw materials into ingredients
    - Manufacture of ingredients into products
- Packaging
- Transportation
- Use and maintenance of the product by the purchaser
- Disposal

By putting these two concepts together, LCA becomes nothing more than measuring and analyzing the environmental effects of the inputs and outputs at each stage of the product life cycle.

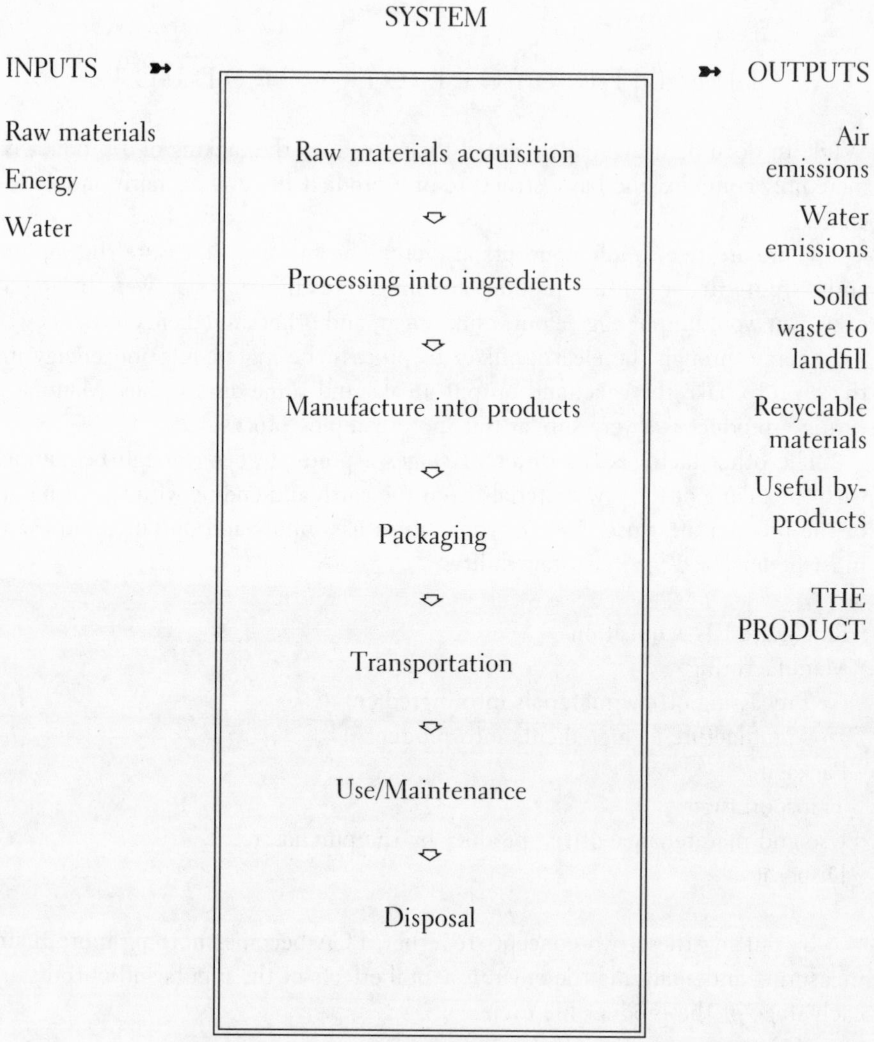

SYSTEM

INPUTS →		→ OUTPUTS
Raw materials	Raw materials acquisition	Air emissions
Energy	⇩	Water emissions
Water	Processing into ingredients	Solid waste to landfill
	⇩	Recyclable materials
	Manufacture into products	Useful by-products
	⇩	
	Packaging	THE PRODUCT
	⇩	
	Transportation	
	⇩	
	Use/Maintenance	
	⇩	
	Disposal	

To determine the sustainability of a product, each and every one of these steps must be evaluated, and for each step there are many subquestions. Right now there are few, if any, products that are completely, 100 percent sustainable in every way, and that 100 percent sustainability may never be achieved. However, what can be achieved are products that are significantly more sustainable for the environment—products that are taking steps in the right direction, products that are doing more good than destruction.

RAW MATERIALS ACQUISITION

This stage of the LCA includes all the activities required to grow, gather, mine, or otherwise obtain or create the raw materials that are used to make the product, and ship it to the processing plant.

The raw materials from which products are made are mined, grown, raised, or harvested by a variety of different methods that either preserve the ecosystem or destroy it, that restore soil fertility or deplete it, that allow for continuous regeneration or not. The raw materials are gathered from the geographical region in which they occur and are transported by fuel-burning, pollution-creating trucks, ships, airplanes, and other methods to locations near and far. With today's global trade, resources are exchanged all over the world, and any given product may contain ingredients from many different countries.

Questions to ask about the raw materials in a product could include:

- Are virgin raw materials used: petrochemicals, metals, minerals, plants, animals?
- Are the resources used renewable or nonrenewable?
- Are recycled materials used: metal, paper, glass, plastic?
- Are reclaimed materials used: wood, paper, cloth?
- How are the raw materials grown, raised, or otherwise obtained? Are standard procedures used, or are special practices used such as organic agriculture, wildcrafting, or sustainable forestry? What kinds of resource inputs (energy, water, fertilizer, etc.) are used to produce the raw materials? (Some of the questions listed under manufacture below may also apply to this step.)

MANUFACTURE OF INGREDIENTS AND PRODUCTS

The manufacturing stage of the LCA has two parts:

- Processing raw materials into ingredients—the activities required to process a raw material into a form that can be used to make the product (or package).
- Product manufacture—the activities required to make the end product that goes in the package.

One Way to Present Product Life Cycle Information
This advertisement for Anheuser-Busch beers is an excellent example of how product life cycle information can be presented for a product. It shows both the environmental attributes in each phase of Anheuser-Busch's production process and clearly states the company's philosophy. With this kind of information, consumers can better evaluate the potential environmental benefits of choosing these beers over other brands. Courtesy of Anheuser-Busch Companies.

Many raw materials go to a processing plant where they undergo interim processing into an "ingredient" that is then used by the manufacturer to make or formulate the product sold on the retailer's shelf. A shampoo manufacturing plant, for example, would not take delivery of crude oil and coconuts to make the ingredient cocomide DEA—it buys the cocomide DEA already manufac-

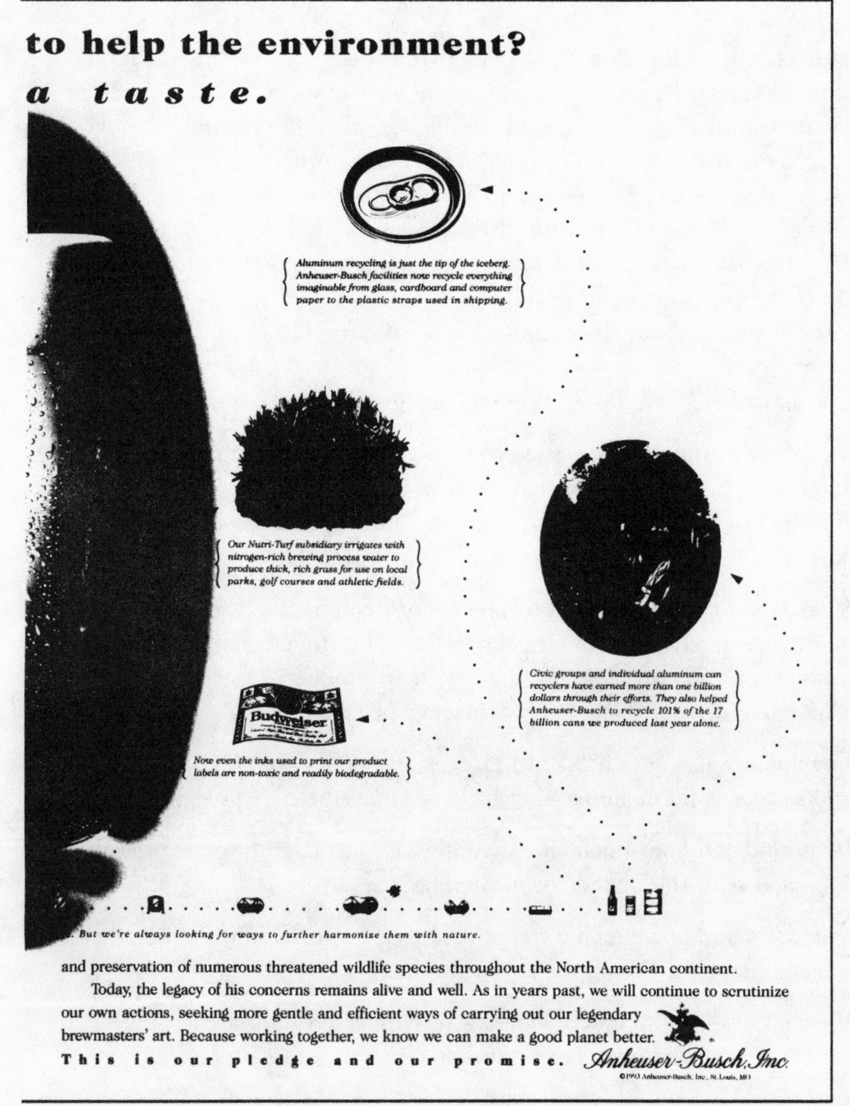

tured, and mixes it with the perfume oils produced by another manufacturer and the artificial colors made by yet another manufacturer. The shampoo manufacturer only knows that the ingredient came from Manufacturer X, and in most cases has no knowledge of the original source of the raw materials.

Some ingredients may undergo several levels of processing before they reach the product manufacturer. Cotton being made into fabric, for example, goes from the farm to the gin to the spinning mill to the weaving mill to the dye house and to the finishing mill before the fabric is shipped to the cut-n-sew factory where the garment is made.

Fuel-burning transportation is again a factor as ingredients travel from manufacturer to manufacturer. Manufacturing processes utilize water, energy, and other resources and create air pollution, water pollution, and solid waste.

Manufacturing is the step about which the least is known to consumers. Often manufacturers themselves are not aware of issues such as how much energy they are using or how much solid waste goes off to the landfill.

One step many manufacturers have taken to lessen their environmental impact is to change to energy-efficient lighting. More than five hundred major manufacturers have signed up for the EPA Green Lights Program in which the EPA offers technical assistance and publicly recognizes corporations that commit to installing energy-saving lighting systems.

Questions to ask about the manufacturing of a product could include:

- How and where were the ingredients processed?
- What type of transport was used to ship ingredients to the manufacturers? What was the distance? What kind of energy was used, how energy-efficient was the vehicle, what quantity and type of air emissions were created?
- What kind of energy is used to process or manufacture the ingredient or product: coal, oil, gas, hydro, nuclear, solar? How much energy does it take to process or manufacture the ingredient or product? Does the processor or manufacturer have an energy-efficiency program?
- How much water does it take to process or manufacture the ingredient or product? Does the manufacturer have a water-efficiency program?
- What kind and how much air emissions are released during the process or manufacture of the ingredient or product?
- What kind and how much water emissions are released during the process or manufacture of the ingredient or product?
- What kind and how much solid waste is released during the process or manufacture of the ingredient or product?
- What kind and how much toxic waste is released during the process or manufacture of the ingredient or product?
- Is there any danger to workers in the process or manufacture of this product? What sort of protection is provided? Are workers paid fairly and are there good working conditions?

Packaging

Technically, placing the product in an appropriate package is part of the manufacturing process, though I consider it separately because the environmental claims of so many products focus only on their packaging.

After the product itself has been made or formulated, it is generally placed in some kind of container and sometimes then has one or more layers of consumer packaging. Often packaged products are shrink-wrapped together with plastic in addition to being placed in a box for shipping.

Packages are not generally made by the product manufacturer, but are shipped in from their own manufacturers. Depending on the product, the product might be packaged in the manufacturer's factory, or shipped to yet another factory for packaging. Each layer of packaging is really a separate product itself, and needs to be evaluated through the same complete life cycle as the product. When there are several parts to a package (say, a plastic holder for the product, with printed instructions, in a box, with a plastic outer wrap), each part of the package needs a separate life cycle assessment.

Packaging really needs to be evaluated separately from the product itself because sustainable products may come in unsustainable packages, and unsustainable products can come in sustainable packages. Some experts say that the environmental effects of the products and package need to be considered as a whole because that is how they are purchased. I agree, but this kind of thinking does not end up encouraging the market for an environmentally exceptional product that might need better packaging. Fortunately, as new products come out that have environmental attributes, having good packaging is an integral part of the product design process.

A lot of attention has been placed on packaging lately because of our garbage crisis. Fifty percent (by volume) of all our trash is packaging. Many states now have packaging restrictions and waste reduction legislation that affects packaging.

The Coalition of Northeastern Governors (CONEG) has developed model state legislation designed to dramatically reduce the amount of packaging entering the solid waste stream. Designed to promote efficient solid waste reduction reuse and recycling programs, the packaging waste reduction model establishes a hierarchy of preferred packaging guidelines beginning with the most preferred:

- No packaging
- Minimal packaging
- Reusable packaging
- Recycled or recyclable packaging.

The best packaging materials are glass, paper and paperboard, steel, aluminum, wood, and packages made of two or more of these components.

There are two types of packages to avoid altogether, if possible. One is virgin plastic packaging that can't be recycled (most food wrap falls into this category, like the plastic bag spaghetti comes in). The other is complicated multimaterial packages that cannot be recycled or reused, and won't biodegrade or burn. The aseptic cartons made of paper, foil, and plastic that are being used for juice, soy milk, and other liquid products are a perfect example (though the aseptic carton industry is trying to set up a recycling program), as are potato chip bags made with plastic and foil.

Transportation

After manufacture and packaging, the product is then transported to the distributor's warehouse (more fuel-burning trucks and ships) and transported again to the retail outlet. Then we drive to the store and buy it, or order it from a mail-order catalog (more fuel, more pollution).

Though I've chosen to put transportation here as a separate issue, it actually needs to be added to every part of the LCA—every time the process of making the product proceeds to the next step. We must consider the transportation of the raw materials to the processing plant, the ingredient to the manufacturer, the product to the warehouse, the retail store, or mail-order catalog and to the consumer, and the waste of the product to the recycling center (and then to the recycling plant) or to the landfill.

Questions to ask about the distribution and transportation of a product could include:

Where was the product (and package) manufactured?

What type of transport was used to ship the products to the retail outlet (or to the next LCA step)? What was the distance? What kind of energy was used, how energy-efficient was the vehicle, what quantity and type of air emissions were created?

Use and Maintenance

This stage of the LCA covers any activity in which the product or package is used, maintained, reconditioned or repaired, or serviced in any way to extend its useful life.

When we use the product there are issues regarding safety to consumer health and resources needed to use and care for the product. How much energy does a light bulb burn? Or how much water flows through a shower head? Is the product toxic? Can it cause accidental poisoning or create pollution? Must it be maintained by using a toxic product (such as dry cleaning only or periodic applications of a toxic floor wax)?

Questions to ask about the use of a product could include:

- Is the product safe for human use (any cautions)?
- Does the product remove pollutants from air or water that might be harmful to health?
- Does the product use or produce renewable, nonpolluting energy?
- Is the product energy- or water-efficient in use (what are the savings)?
- Are air emissions produced by use? What kind and how much?
- Are water emissions produced by use? What kind and how much?
- Is solid waste produced by use? What kind and how much?
- Is toxic waste produced by use? What kind and how much?
- What is needed to care for or maintain the product?

Disposal

At the end of its life a product is thrown away and either ends up in a landfill, is recycled, or biodegrades back into the natural cycle of life.

Questions to ask about the disposal of a product could include:

Is the product reusable (not disposable after one or a few uses)?

Is the product recyclable? Are local recycling facilities for this material available?

Is the product biodegradable? Are all the ingredients biodegradable? Will the product degrade in aerobic and anaerobic conditions, in a landfill or compost pile? How long does it take to biodegrade? What does the product (or each ingredient) break down into? Is the product or the substances that it breaks down into harmful to the environment in which it breaks down?

A LIFE CYCLE ASSESSMENT FOR A REAL PRODUCT

Manufacturers have been doing life cycle assessments of sorts on their products for years, even though they do not generally release this information to the public. One LCA I was able to obtain from a manufacturer is from Church & Dwight Company for **Arm & Hammer Baking Soda,** done by Arthur D. Little, Inc. Though it doesn't address the mining of the ore, by looking at this life cycle assessment, you can see how the product is relatively environmentally benign in its manufacture, use, and disposal. Here are some excerpts taken from a document many pages long.

THE BAKING SODA LIFE CYCLE

Product Description: Arm & Hammer baking soda is 100 percent pure sodium bicarbonate ($NaHCO_3$). It is a natural inorganic sodium salt, derived from trona ore mined in southwestern Wyoming. Baking soda is a white crystalline powder with no perceptible odor and a slightly salty taste. It is mildly alkaline in solution and a readily available source of carbon dioxide. When heated or . . . in acidic conditions, it will give off carbon dioxide. Baking soda is of low toxicity and is not classified as a health hazard; it is generally recognized as safe for humans and animals.

Raw Material Acquisition and Material Manufacture: Arm & Hammer baking soda is made from soda ash (which is derived from a naturally occurring ore called trona), carbon dioxide, and water. Trona ore consists of 86 to 95 percent sodium sesquicarbonate ($Na_2CO_3.NaHCO_3.2H_20$), 5 to 12 percent clays and other insoluble impurities, and water . . . The study focused on the transformation of trona to soda ash and then to baking soda.

Soda ash, or sodium carbonate (Na_2CO_3) is most often produced in the United States from trona ore using the monohydrate process. The mined trona ore is crushed, screened and heated in a calciner, so that water and carbon dioxide are driven off, leaving behind crude sodium carbonate. The crude sodium carbonate is then fed into leach tanks, where it dissolves. The insoluble impurities are separated out through a series of filtration steps. Next, sodium carbonate monohydrate ($Na_2CO_3.H_2O$) is crystallized out from the purified liquid. The final step is drying, where the free and chemically bound water is evaporated from the sodium carbonate monohydrate, resulting in the final product.

Carbon dioxide, the other principal raw material used in Arm & Hammer baking soda, is produced from the gases given off by other commercial manufacturing operations . . .

Product Manufacture: Baking soda is manufactured by dissolving soda ash in purified water, filtering the solution to remove insoluble materials, and then adding carbon dioxide to the purified soda ash solution. Sodium bicarbonate crystallizes to form a slurry in this process. Centrifuging concentrates and removes the bicarbonate solids, which are dried and cooled. The remaining centrate is recycled to the beginning of the process for use in the soda ash dissolution step.

Packaging: Arm & Hammer baking soda is packaged for use by households in cartons made of recycled clay coated newsback board (primary packaging). The cartons are made from approximately 57 percent post-consumer material, such as municipal waste newspaper, corrugated boxboard, and computer printout paper . . . They are made from approximately 38 percent pre-consumer material, such as material from paper mill printing and cutting conversions. They also contain about 5 percent coating material. Inks used for packaging are food-safe . . .

Use and Reuse: Contrary to the single purpose suggestion behind its name, baking soda is actually a multi-purpose product [uses listed are baking ingredient, bath additive, cat litter deodorant, dentifrice, drain freshener, laundry additive, refrigerator/freezer deodorant, surface cleaner (kitchens and bathrooms), and swimming pool pH adjuster].

In all uses considered except one, baking soda is used once and then discarded [after being used as a refrigerator/freezer deodorant it can then be used as a drain freshener]...

Transportation: The environmental loadings from transportation of raw materials, manufactured products and wastes were considered at each step of the baking soda product life cycle. Transportation stages include vehicular activity that is a part of trona mining and benefication operations, transport of raw materials (soda ash and carbon dioxide) to the baking soda manufacturing facilities, transport of packaged baking soda to stores/households, and transport of manufacturing, distribution and post-consumer wastes to landfills and incineration facilities.

Fuel consumption in energy equivalents (million BTUs) and air emissions were calculated for each of the life cycle stages outlined above and then aggregated for each of the nine uses. Energy and air emissions loadings associated with vehicular activity that is part of trona mining and benefication operations were reported with other environmental loadings associated with raw material acquisition and material manufacture, respectively.

Waste Management: Wastes that are generated during the baking soda product life cycle include tailings from the trona benefication process, baking soda manufacturing wastes, baking soda disposed of by consumers, baking soda packaging, and ash from incineration ...

Tailings from the trona benefication process typically are made up of shale, natural clay and natural silica. They are the impurities in trona ore—the components that are not water soluble. Tailings are pumped from the soda ash production process to large ponds, where they are dewatered forming nonhazardous solid waste. Baking soda manufacturing wastes are landfilled ...

Baking soda that is not consumed or dissolved in water during household use is usually disposed of in household trash ... As part of household solid waste, baking soda is typically collected at curb-side and transported to either landfills or incinerators.

Baking soda that is in solution at the end of a particular use is usually disposed of down drains connected to sewage systems that flow into treatment plants. In the process of dissolution, the baking soda dissociates into sodium ions and bicarbonate ions. Placement of baking soda in drains contributes to the alkalinity of waste water, and may result in less alkali being added at waste water treatment plants.

Primary packaging is generally disposed of as household solid waste. Although the board fibers are technically recyclable, the recycling rate for these cartons is insignificant ...

Results of the Life Cycle Inventory

Waste Water: The data used for this study show no adverse waste water releases associated with the nine household uses ...

Air Pollution: From the standpoint of public environmental concerns, atmospheric emissions of carbon dioxide pose the most significant environmental loadings associated with the baking soda life cycle. Carbon dioxide is a greenhouse gas ... All air

ARM & HAMMER BAKING SODA LIFE CYCLE INVENTORY

Per ton baking soda used:

Raw Materials (lbs.)	**2,999.000**
Product Input	2,836.000
Packaging	163.000
Water Consumption (gals.)	**163.000**
Energy (MM BTU)	**5.820**
Diesel	1.509
Electricity	2.169
Unleaded Gasoline	0.001
Natural Gas	2.217
Energy Credit (Packaging)	−0.075
Atmospheric Emissions (lbs.)	**1406.447**
Carbon Dioxide	1,384.000
Carbon Monoxide	2.059
Nitrogen Oxide	5.056
Sulfur Oxide	3.654
Particulate Matter	1.202
Methane	10.000
Unspecified Hydrocarbons	0.450
Unspecified VOCs	0.026
Waste Water Pollutants (lbs.)	**0**
Hazardous Waste (lbs.)	**0**
Solid Waste (lbs.)	**1,217.000**
Process	1,064.000
Nonprocess Manufacturing	16.000
Post-User	137.000

emissions studied except carbon dioxide average less than 10 pounds per ton baking soda . . . Most of the carbon dioxide emissions are associated with energy requirements for [manufacturing and transportation].

Solid Waste: . . . With respect to solid waste, the route of disposal is significant. Disposal down the drain contributes to fewer environmental loadings than disposal in the trashcan. Fully 80 percent of baking soda used by households is disposed of down the drain . . . The baking soda life cycle is not associated with any significant generation of hazardous waste.

Conclusions

The results of this study indicate that, on the whole, baking soda is environmentally benign.

If we had information like this on every product, we, as consumers, could begin to compare the environmental impact of our product choices. It, of course, will not fit on the label, but should be available by request.

SCS ENVIRONMENTAL REPORT CARD

While many manufacturers use some form of LCA to analyze or make improvements in their own product manufacture, Scientific Certification Systems (SCS) has developed a simplified life cycle assessment for manufacturers to use in communicating this information to consumers: the Environmental Report Card certification program.

Called "the environmental equivalent of a nutritional label" by SCS, the Environmental Report Card gives a detailed accounting of the environmental burdens (see sidebar) associated with the product and its packaging, just as nutritional labels state how many calories and how much fat and vitamins are in a product. These resource inputs and waste outputs are tracked from the time raw materials are extracted from the ground through the manufacture, transportation, use, and disposal of the product. The life cycle totals for each burden are then presented on the label both in numerical quantities and plotted on a bar graph.

The advantage of such a labeling system is that, like nutritional labels, there is no particular judgment of a product being good or bad for the environment—the label simply shows the numbers, and allows us to make our own decisions, based on accurate, unbiased facts. If the government wants to regulate something, I personally would like to see it require this kind of information on every single product, as well as complete ingredient listings, just as it requires nutritional information.

SCS acknowledges that this method is not perfect (and, indeed, there may never be a perfect system that accounts for the complexity of every aspect of every product). However the Environmental Report Card gives us more infor-

SCS ENVIRONMENTAL BURDENS

All products place some burden on the enviroment from resource use and waste disposal. Products that place less burden on the environment are more sustainable; products that place more of a burden on the environment are less sustainable. By knowing the quantitative burdens of products, you can compare them and choose the most sustainable. The SCS Environmental Report Card evaluates enviromental burdens for products in the following five categories:

	PRIMARY ENVIRONMENTAL CONCERN
Resource Depletion	
Water (Fresh)	Depletes fresh water; degrades wildlife habitats
Wood	Depletes forests; degrades wildlife habitats
Coal, Oil, Natural Gas (nonfuel)	Depletes nonrenewable resources; degrades wildlife habitats
Minerals	Depletes nonrenewable resources; degrades wildlife habitats
Food/Fiber (e.g., cotton)	Land degradation resulting from nonsustainable farming practices
Animal Products (e.g., leather)	Land degradation resulting from nonsustainable animal production practices
Soil	Topsoil erosion
Energy Use	
Total Energy Used	Depletes nonrenewable resources; degrades widlife habitats
Coal	Substantial air emissions, water emissions, and solid waste
Oil	Substantial air emissions, water emissions, and solid waste

Natural Gas	Substantial air emissions, water emissions, and solid waste
Nuclear	Radioactive, long-lived solid waste
Hydro	Degrades wildlife habitat
Air Pollution	
Ozone Depletors	Increased skin cancer, cataracts; disrupts food chain
Carbon Dioxide	Global climate changes
Carbon Monoxide	Smog; asthma; lung cancer
Sulfur Oxides	Acid rain, leading to destruction of lakes, crops, forests
Nitrogen Oxides	Acid rain, leading to destruction of lakes, crops, forests; smog
Particulates (dust)	Disrupts plant growth; human and animal respiratory problems
Unclassified Chemicals	Unknown hazards to humans, plants, and animals
Hazardous Air Pollutants	Smog; asthma; lung cancer; stunts plant growth
Water Pollution	
Total Solids	Degrades water quality; degrades aquatic wildlife habitats
Oxygen Depleting Chemicals	Depletes oxygen in water, suffocating aquatic life (measured as BOD—Biochemical Oxygen Demand)
Toxic Pollutants	Cancers; mutations; reproductive damage; contamination of food supply
Solid Waste	
Hazardous Waste	Potentially contaminates groundwater, rivers, lakes, and oceans
Unclassified Waste	Contributes to landfill problem; pollution from incineration

Source: Scientific Certification Systems.

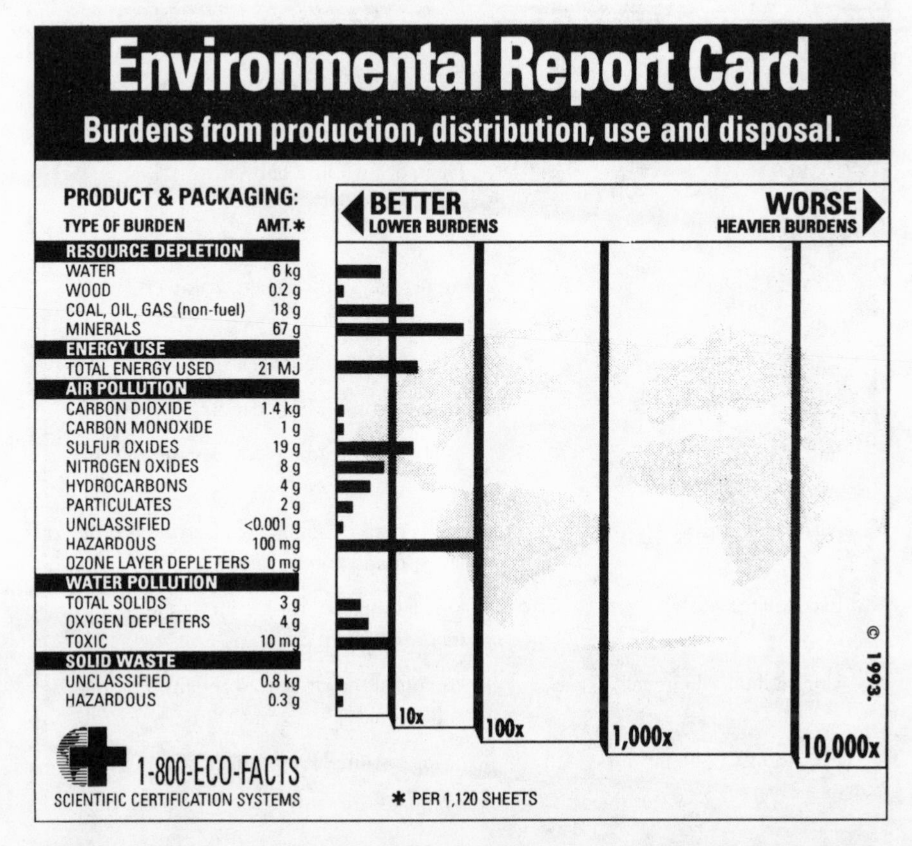

TREE-FREE 100% RECYCLED 2-PLY BATH TISSUE:
(SIX ROLLS OF 600-SHEETS)

The SCS Environmental Report Card gives a two-part summary of the environmental impact of a particular product. In addition to providing a quantitative breakdown of environmental burdens, it is designed to give consumers an easy-to-understand graphic depiction of how well a product's impact compared to similar products.

mation in a concise, easy to compare format, than any other method currently available. Individual environmental attribute claims can give us some general idea of improvements (for example, in general, recycled paper saves trees, energy, and water), but there's no comparison like looking at the hard facts about individual products (see sidebar).

Only a handful of companies currently have Environmental Report Cards for their products, but it seems likely that such labeling will become commonplace as data collection methods are perfected and consumers show interest in choosing products on the basis of such information.

EVERY PRODUCT HAS A STORY

In addition to describing a product as a scientific tally of environmental sources and sinks, the life cycle of a product can also be told as a story. It can tell not only the physical characteristics of the product, but also about the people who make the product and their attitudes, hopes, and dreams. There may be an interesting historical aspect to a product, or special practices used that have value beyond a simple measurement of inputs and outputs.

The pertinent life cycle story of a product can be told quite briefly, as in this description I wrote for the label of **Barbecubes Natural Fruitwood Briquets:**

> THE BEST BRIQUET FOR THE ENVIRONMENT
>
> Barbecube is one of the few products on the market today that offers environmental benefits in every step of its manufacture, use, and disposal.
>
> MADE FROM CERTIFIED RECLAIMED FRUIT TREE PRUNINGS
>
> Each autumn, at the end of every harvest, fruit trees are routinely pruned to encourage abundant fruiting the following season. Standard practice is to burn the prunings in the field, which wastes this valuable source of energy and causes air pollution. These prunings are collected to make Barbecubes.
>
> PROCESSED WITH WASTE HEAT
>
> The collected prunings are taken to a nearby processing plant, where they are chipped and dried using waste heat from a co-generation plant that produces electricity for the region.
>
> DRIED WOOD BURNS CLEAN
>
> Complete combustion of dried wood produces only carbon dioxide and water, the same elements that are released when a tree decays naturally in a forest. Drying the wood greatly reduces the air pollution caused by burning the wood green in the field. And burning wood causes much less pollution than burning charcoal.
>
> ASH NOURISHES YOUR GARDEN
>
> Barbecube briquets leave only a fraction of their volume in natural ash, which can be used to return vital minerals to the soil in your garden. Ash from charcoal briquets cannot be used in the garden.
>
> A TREE IS PLANTED FOR EVERY BAG SOLD
>
> Return the coupon, and Barbecube will plant a tree in your name by the Global ReLeaf Heritage Forests program. This helps the American Forestry Association to create new forest habitats on public lands that would not be possible under existing public programs.

INDUSTRIAL ECOLOGY

Beyond looking simply at the life cycle of a product in the closed system of its manufacture, the next step is to see how companies can and do interconnect with other companies for the symbiotic sustainability of both. The emerging field of industrial ecology offers a large-scale, integrated management tool for designing industrial infrastructures to be interlocking artificial ecosystems interfacing with each other and the natural global ecosystem. In essence, the manufacturing processes are designed into the existing natural processes and cooperate with each other in the exchange of resources, much the way all the elements of nature interact to create a thriving and productive system.

While this idea is, for the most part, still on the drawing board and not yet widely applied in industry, there are some examples of companies that have already begun to think along these lines. The **Barbecubes Natural Fruitwood Briquets** mentioned earlier are a good example of industrial ecology on a small scale. The briquets themselves are made from a waste product of nearby orchards that would otherwise have been burned green, creating air pollution (if you've never been in an agricultural area at the time of year when they are burning prunings and other dry waste agricultural materials, take my word for the fact that the smoke is so thick for miles that you can hardly breathe). The heat used to dry the wood is the waste heat from the neighboring power plant that generates electricity, thereby making the fuel utilization more efficient. Barbecubes even has a useful link with the customer, as the ash can be used to fertilize the garden. Such cooperative interconnections should be standard practice in business, and probably will be as companies recognize this is both practical and profitable.

USING LIFE CYCLE TO EVALUATE PRODUCTS

Don't be discouraged if this all sounds complex. I don't expect any consumer to write a detailed life cycle assessment, but I've included this chapter so you will have a better idea of why it is so difficult to evaluate the sustainability of a product. There are a mind-boggling number of details to consider, and those details are specific to each manufacturer.

Being aware of the basics of life cycle considerations when you shop, however, can help form a more complete picture of the environmental pros and cons of a product.

Start by looking at the ingredients, then see if there is anything known about the manufacturing practices. Note how far the product was shipped to reach your store. Does the product protect your health because it is nontoxic, or does it protect the environment by saving energy or water during use? And finally consider how you will dispose of the product and package after use.

CHAPTER 4

Renewable and Natural

The most basic question in determining the sustainability of a product is to evaluate the natural resources from which it is made. Every product we buy or use is made from the natural resources of the earth. Every product utilizes the basic raw materials from which it is made, plus the energy and water resources needed in its manufacture. Resources are used to build the factories to make the product, the trucks and boats and planes in which the product is transported, the stores in which the product is purchased, and the car we drive to the store. Energy resources are burned in the factory, in transportation, and in the retail store.

We don't generally think of products as being resources transformed. It's common to have the limited impression that the products we buy come from the store or maybe the factory, but in reality, all products begin with natural resources from the earth—crude oil, metals and other minerals, trees and other plants, animals, and water. When I started to see every product on the shelf as piles of the raw resources from which they were made, I began thinking about them differently and had more appreciation for what went into them, rather than of thinking of a product as just a commodity. Instead of thoughtlessly buying paper towels, I imagined them as the forest of trees that were clear-cut so they could be made, or huge piles of paper collected at a recycling center. Thinking this way about products will change your perception of what a product is, and how it is connected to the earth.

A resource can be defined as "a source of supply . . . especially one held in reserve," "the collective wealth of a country or its means of producing wealth," and "money, or any property that can be converted into money, assets"—all consumer culture definitions referring to material that can be turned into something valuable. In the consumer world a resource is something that has little value for its own sake, but much value when it has been made into something else; a tree has value only if it can be made into lumber, land has value only if it can be farmed or we can build houses on it.

But the root of the word *resource* is the Latin *resurgere,* "to rise up again." In this sense, a resource is a constant, ever-renewing, sustainable supply; much like, as Wendell Berry says, "a spring refills its basin after a bucket of water has been dipped out." With the right sustainable use—respecting the source, respecting the ability of the source to resurge, and not asking the source to provide more than is natural for it—resources naturally replenish themselves and can last as long as the earth. Such a resource is sustainable and has value in its own right, for its function in nature; trees have value to us in providing shade, regulating climate, and circulating water through the systems of the earth, wilderness land is priceless in its beauty and inspiration.

Resources used to make products are classified as being renewable or nonrenewable. In order for any product to be sustainable, it must begin by being made from resources that can reasonably be restored and replenished by nature. Of course, all the resources of the earth are constantly being made by nature, but there is an immense difference in the time it takes for a microorganism to reproduce, a fish to be born, a tree to grow, and crude oil to form. Whether a resource is renewable or nonrenewable is not so much about whether or not nature has the ability to make more of something, but about how long nature takes to create it. Ideally, we should use all resources at the rate at which they can be replenished by nature. Nonrenewable resources are considered nonrenewable because we use them much faster than they can be renewed for us to use again.

NONRENEWABLE RESOURCES

Nonrenewable resources are those which cannot be replenished by nature within our lifetimes, and are therefore in limited supply. Nonrenewable resources are fossil fuels, metals, other minerals, and groundwater.

Since nonrenewable resources cannot replenish themselves on any time scale that is relevant to our lives, nonrenewable resources are inherently not sustainable. If we were to use these resources at the rate at which nature replenishes them, they would be so valuable as to be priceless because of their scarcity. They are not easily replaceable and are environmentally damaging to acquire, yet right now our entire society and economy are based on the use of nonrenewable resources, particularly fossil fuels.

While more than 90 percent of the fossil fuels mined are used for fuel, virtually every product we use is made from substances derived from fossil fuels or utilizes fossil fuels in its manufacture or transportation. Common products made from fossil fuels include detergents, synthetic fibers (nylon, polyester, acrylic), plastics, paints, garden hoses, food additives, cleaning products, pesticides, nail polish, lipstick, shampoo—every type of product on the market would have to be listed here, for if it is not made purely from petroleum products, some petroleum derivative is used in the making or packaging of the product, and in its transport to various markets around the country and around the world. Of course, there are exceptions. I can imagine a baker walking out into his field, cutting some wheat, hand-grinding it, baking the loaf in a wood-fired brick oven, and having the customer walk to the bakery to buy it—but this is not commonly how it's done.

Fossil fuels are not a sustainable resource: They are in limited supply and cannot be replenished, they create pollution that destroys ecosystems, and they are harmful to health. As we empty our reserves for fuel and consumer products, the fossil fuels become pollution in our water, land, and air—they don't ever come together again to make more fossil fuels.

All the fossil fuels—crude oil, coal, and natural gas—are made by nature from plant material by very similar processes. Through photosynthesis, plants use energy from the sun, carbon dioxide from the air, minerals from the soil, and water to produce glucose, a kind of sugar. Glucose and the substances derived from it are all compounds of carbon. So the fossil fuels are really no more than processed plants that contain stored solar energy, which we now are using millions of years later.

Our extensive use of fossil fuels is very recent. The first commercial uses of crude oil, coal, and natural gas were as fuel. The first coal mine in America was opened in Virginia in the 1700s, the first commercial oil well was drilled in 1859 in Pennsylvania, and the first well for natural gas was drilled in New York in the 1820s. In less than two hundred years fossil fuels and their derivatives have become the lifeblood of our modern technological world. Our agriculture, industry, transportation, and communication systems depend on a steady and economical supply of fossil fuels—without it our way of living would be very different. Yet for most of our history and before, our species survived fairly well without them.

The discovery, manufacture, and distribution of fossil fuels and their products is highly technological, consuming fossil fuel resources and creating fossil fuel wastes every step of the way. We are all aware of large oil spills from leaking tankers, but it is not generally known that small oil spills happen on a routine basis as a result of accidents to drilling rigs, pipelines, storage tanks, and tankers. On the North Slope of Alaska alone, there are between four hundred and six hundred reported oil spills each year—more than one every day. In total, about .1 percent of the world's oil production, some 1.4 *billion* gallons, ends up polluting oceans each year, killing marine life, damaging the spawning grounds of fish and other aquatic animals, and contaminating our food supply. In contrast, the devastating *Exxon Valdez* oil spill was only 11 *million* gallons. According to The Wilderness Society, oil spills occur, on average, twenty-five to thirty times each day. During the one-year period between March 1990 (the one-year anniversary of the *Exxon Valdez*) and mid-March 1991, the fifty worst spills totaled 13.4 million gallons.

Here in California, we have a law that requires the Governor of California to publish a list of chemicals "known to the State to cause cancer or reproductive toxicity." The law also requires that a clear and reasonable warning be given to citizens exposed to the listed chemicals in particular situations. On November 17, 1992, the *San Francisco Chronicle* published an advertisement that read:

> WARNING
>
> Chemicals known to the State to cause cancer, birth defects, or other reproductive harm are found in gasoline, crude oil, and many other petroleum products and their vapors, or result from their use . . .
>
> Chemicals known to the State to cause cancer, birth defects, or other reproductive harm are found in and around gasoline stations, refineries, chemical plants, and other facilities that produce, handle, transport, store, or sell crude oil and petroleum and chemical products.
>
> Other facilities covered by this warning include, for example, oil and gas wells, oil and gas treating plants, petroleum and chemical storage tanks, pipeline systems, marine vessels and barges, tank trucks and tank cars, loading and unloading facilities, and refueling facilities.
>
> [signed] ARCO (Atlantic Richfield Company), BP Oil, Chevron, Exxon, Mobil Oil, Phillips Petroleum, Shell Oil, Texaco, UNOCAL, Valvoline [and other oil and gas companies].

Whether or not we choose to change our use of fossil fuels for environmental or health reasons, sooner or later we will not have a choice because we are simply running out of these nonrenewable resources.

The United States has been the world's biggest oil producer for almost a century. We were self-sufficient in petroleum—producing enough for our own needs and exporting to other countries—until 1948, when we began to import more oil than we exported. After the 1973 Arab oil embargo that raised prices for foreign oil, we began to have increasing interest in domestic production.

In 1988, 24 percent of world oil production came from the Middle East and 21 percent from the USSR; these two regions between them have 72 percent of the known oil reserves and 40 percent of estimated undiscovered reserves. This means that the United States has less than 28 percent of known reserves. Already we have depleted much of our easily available domestic crude oil, and now rely on supplies shipped in from Alaska or other countries. We can't go on like this for much longer. At our current rate of usage, we have enough crude oil and natural gas worldwide to last only about fifty years, with domestic sources running dry in thirty years. Clearly, the ability of the United States to continue to power itself on oil and make everyday products from petrochemical derivatives is on the decline.

The other mineral mainstay of our technological world is metals. Metals are limited, precious gifts from the earth, used since the earliest days of human history. Like the fossil fuels, they are formed by geologic processes of tremendous force over millions of years, and mining them is destructive to the environment. Hundreds of thousands of acres of land and miles of waterways have been polluted by mine drainage and sediment, and smelter effluents, which are extremely toxic. Surface, or strip, mining, in which huge power shovels remove rock, vegetation, and soil covering metals and other minerals, has destroyed the natural habitat of more than 2 million acres of land in the United States alone. Regulations now require land restoration after strip mining, but even the best efforts cannot restore the land to its original state, and about two thirds of strip-mined areas still remain wasteland.

As we use metals, their nonrenewable source becomes more and more depleted, as already evidenced by the necessity to mine lower and lower grade ores. Eventually, though some metal will still be present, we will reach a point where it is neither economically nor environmentally feasible to mine the lowest grade ores as they require more fossil fuels to obtain and create more waste in mining. Since surface deposits are no longer sufficient to fill the needs for certain minerals and metals, mining has become a complicated, expensive, extremely energy-intensive, and highly technical task.

Even though they are not sustainable by nature, nonrenewable resources can be used in a somewhat sustainable way by using them at the rate at which renewable resources, used sustainably, can be substituted for them. An oil deposit, for example, could be used sustainably if part of the profits from it were systematically invested in solar collectors or in tree planting, so that when the oil is gone, an equivalent supply of renewable energy is still available.

Another way a nonrenewable resource can be sustainable is through recycling. We have billions of tons of metals that have already been mined and could be recycled (many already are). Many crude-oil products, such as plastics and solvents, could be recycled at a greater rate than they currently are.

RENEWABLE RESOURCES

Renewable resources are those which can be replenished over and over again within an average human lifetime and have the potential to be virtually inexhaustible, if they are managed sustainably. Plants, animals, surface water, soil, and sunshine are all renewable resources. I also include certain basic minerals (such as salt or clay or the silica used to make glass) in my definition of renewable because they are among the basic building blocks of nature and are extremely abundant.

For a renewable resource to be sustainable, the rate of use can be no greater than the rate of regeneration. Thus, for example, fish are harvested sustainably when they are caught at a rate that can be replaced by the remaining fish population, and trees are harvested sustainably when the number of trees taken is less than or equal to the number of new trees that grow. Surface water is sustainable when the amount used is less than or equal to the amount of rain or snowfall that replenishes it.

That a resource is renewable is a prerequisite to sustainability, but it is only the first step. Once you have determined that a product is made from a renewable resource, you also have to look for some indicator of sustainable practices. Look for products that are organically grown (see Chapter 5), recycled (see Chapter 6), or sustainably harvested (see Chapter 7). Agriculture, for example, is sustainable when the fertility of the soil is maintained, so that crops can continually be planted and harvested without depleting the vitality of the field. By contrast, agricultural methods that employ toxic pesticides and artificial fertilizers destroy topsoil and eventually make it impossible to use the land for growing crops. Even though food plants are a renewable resource, poor farming practices can reduce or even eliminate supply.

I learned a lesson in sustainable use of a renewable resource from the deer in my backyard. When we were starting to plant our vegetable garden, we planted some little strawberry plants before putting up the deer fence. The first night the deer came down into the garden while we were sleeping and ate most of the leaves off the strawberries. I was so upset, thinking that they had killed the plants, but there were a few leaves remaining and we kept watering the plants, and now they have grown to be lush and full of strawberries. I then noticed another plant, a native similar to the strawberry in that it has single leaves at the end of long stems, that had been eaten by the deer and it was coming back, too. Since that time I've watched them graze: They only nibble on part of the plant, always leaving enough for the plant to survive and regrow. The deer know that if they want to continue to have a food supply, they can only take as much as can be taken and still allow for nature's regenerative processes to restore their munching grounds (and they leave some "fertilizer" behind, too).

I have not found any guidelines or regulations regarding the use of the word *renewable* on product labels, and I have only seen it on a very few paper and wood products.

Renewable Products

Many household products have some version available made from renewable resources, or could, if there were sufficient consumer demand. The exception is machinery—telephones, refrigerators, calculators, computers, and the like. But for everyday sustenance products—**FOOD, TEXTILES, CLEANING PRODUCTS, PEST CONTROLS, BEAUTY AND HYGIENE**—choose renewable ones. Creatures in their natural habitat consume only renewable resources—water, insects, fungi, berries, seeds, etc.—we are the only species who mines nonrenewable resources from beneath the earth's crust.

In general, the most sustainable choices are products made from renewable resources rather than those made from nonrenewable resources. Even if the renewable resource is not sustainably managed, I believe it is preferable to buy a product made from a renewable resource (such as a cotton shirt) over one made from a nonrenewable resource (such as a polyester shirt), simply because the renewable resource has the potential to be managed sustainably, and the nonrenewable resources do not (with the exception of those that are recycled).

Some experts disagree. They maintain that given our current resource and manufacturing practices, the harm that comes to the earth and the depletion of resources is about the same, whether we use renewable or nonrenewable resources. They also argue that with our population numbers we have to use nonrenewable resources because renewable resources are "too inefficient" to supply everyone with certain products. It takes less energy and fewer raw materials to make polystyrene coffee cups, for example, than paper cups. Similarly, it would take many acres of land to grow the plant materials needed to make natural fabric dyes if they weren't made from crude oil.

As long as nonrenewable resources are being mined for fuel, it probably makes sense environmentally to use the leftover by-products for consumer products instead of disposing of them as waste—if industry can do so in a way that is safe for all living things.

But the shortsightedness of this kind of thinking is that it is not sustainable—we will one day run out of crude oil. The population of our species should—like every other species—be limited to the renewable resources available for its support. By using nonrenewable resources, we are artificially inflating our population to levels that cannot be sustained by the natural systems of this world. Some say our industrial technology and population have grown too large to ever go back to relying on natural materials. But there is no choice—we are using our nonrenewable resources faster than nature can replenish them.

It is not too soon for us to begin a transition to replacing oil used for fuel and consumer goods with fuel and consumer goods made from renewable, sustainable resources. This doesn't necessarily mean eliminating the use of nonrenewables altogether, but instead using them judiciously where they would provide the most benefit.

NATURAL

If sustaining the earth means acting in harmony with nature, then it only follows that a sustainable product would be a natural product. But when we buy natural foods, natural cosmetics, natural fibers, or any other so-labeled "natural" products, we aren't necessarily purchasing a product that is also sustainable.

The word *natural*, as it relates to consumer products, is meaningless, since every consumer product is made from the natural resources of the earth. Because, as yet, there is no legal definition, natural has been both overused and misused in labels on many products.

Natural is commonly used to mean primarily made of renewable resources, as opposed to man-made ingredients derived from nonrenewable resources. The basic substance or material used to make the product is found in nature (instead of being manufactured from petroleum) and is therefore thought to be more compatible with the human body and the entire ecological system. Materials generally accepted as being natural are those plants, animals, and minerals as they appear in nature (such as cotton, wool, or salt), or ingredients derived from plants, animals, and minerals (such as lavender oil, gelatin, and baking soda).

The main drawback to the term *natural* is that it gives the illusion that the product is "of nature," and therefore absolutely acceptable and harmless to our health and the environment. In fact, for the most part, "natural" products are natural only in the respect that some part of the product exists in nature and it is not completely made from petrochemicals.

Natural products are rarely completely natural—in our modern world of processing, petrochemicals are almost always used with the natural materials when making a "natural ingredient." More accurately, most "natural" substances and materials might be called (I'm coining a new word here) "hybrid-natural"—basically of nature, but grown or processed by industry with added petrochemicals. Virtually all products being currently marketed as "natural" are really hybrid-natural.

One excellent example of a hybrid-natural ingredient was described by Philip Dickey of the Washington Toxics Coalition in an article for *Green Alternatives* magazine (May/June 1992). He specifically wrote of a coconut oil surfactant used in natural cleaning products, but the same principle applies to all so-called natural ingredients.

> This detergent is called linear alcohol ethoxylate . . . The alcohol from which this surfactant is made is similar to the ethyl alcohol we know from beer and wine except it has more carbon atoms . . . They are arranged in a straight line, hence the term

> *linear alcohol.* This alcohol (called lauryl alcohol) can be manufactured from either coconut oil or petroleum. Let's pretend that our alcohol came originally from coconuts. . .
>
> Lauryl alcohol is not a surfactant. To make it function as a surfactant, we have to build on a hydrophilic, or water-soluble, structure . . . in an ethoxylated alcohol it is done through a chemical reaction with a highly toxic (and carcinogenic) compound called ethylene oxide, distilled from crude oil. During this process, called ethoxylation, carbon atoms from ethylene oxide are progressively added to one end of the coconut-based structure until a hydrophilic chain of the desired length is reached. At this point *the surfactant can be thought of as part vegetable, part petroleum . . . a hybrid.* The ratio varies, but is often near 50:50. [italics mine]

Hybrid-natural refers not just to formulated ingredients. The process of making "natural" apple juice (generally accepted to mean being preservative-free, though still made from apples grown with pesticides and other petrochemicals) is anything but natural. After apples are harvested, they are trucked to a nearby processing plant, where they are run through a trough of chlorinated water to wash off dirt, pesticide residues, and other undesirables. Some plants also run the apples through a bath of carnauba wax (also used in car wax) and preservatives. Before the apples become juice, they are peeled by running them through a "bath" of hot sodium hydroxide (lye), which penetrates beneath the skin and loosens it from the pulp. High-velocity jets of cold water then blast off the skin and a machine punches out the cores. The peeled, cored apples are crushed in a high-pressure juicer. To keep the juice from turning too dark, some processors add sulfiting agents, which inhibit the browning enzyme (if sulfites are added, they must be noted on the label). Then it is pasteurized with heat (from 200 to 400 degrees for four or five seconds) to kill any bacteria that might be present, which also reduces the vitamin content. Hardly as "natural" as biting into a crunchy organically grown apple or running the apple through a juicer at home.

Of course, there are some ingredients that appear in products just the way Mother Nature intended, such as organically grown olive oil in a salad dressing or wildcrafted herbs in a salve. But beware of entire products called "natural"—don't take for granted that they are entirely safe for the environment or sustainable in every way.

Natural Products

Even though the word *natural* can be misleading on a label, in a world where much of what we use in our daily lives is made totally from petrochemicals, it can alert us to products that at least have some root in the renewable, potentially sustainable, resources of the earth. All products made from natural ingredients use plants, animals, and minerals (in their natural state, such as salt) as their primary resources. In the marketplace, look for natural foods (see **FOOD**), natural fibers (see **TEXTILES**), natural cosmetics (see **BEAUTY AND HYGIENE**), natural **CLEANING PRODUCTS**, natural **PEST CONTROLS**, and natural **ART SUPPLIES**.

CHAPTER 5

Organic
Organically Grown
Biodynamically Grown
Wildcrafted

One of the key words to look for that indicates a product is made from renewable resources grown in a sustainable way is *organic,* or more correct, *organically grown.* But it's not so simple as it sounds. A laundry detergent has "organic" cleaning agents. An herbal flea collar claims to be "nontoxic" and "organic," with no additional information to define the meanings of these terms. A bottle of apple juice is made with "organic" apples and a clothing manufacturer sells T-shirts made with "organic" cotton. On each of these labels, the word *organic* means something different.

Among the many definitions for organic: "noting or pertaining to a class of compounds that formerly comprised only those existing in or derived from plants and animals, but that now includes all other compounds of carbon"; "characteristic of, pertaining to, or derived from living organisms"; and "a substance, as a fertilizer or pesticide, of animal or vegetable origin." All these definitions are used in some way or another on product labels.

An "organic" cleaning agent is likely to be one of that "class of compounds that formerly comprised only those existing in or derived from plants and animals, but that now includes all other compounds of carbon." Since crude oil and coal both started out as plants and animals and are therefore made of carbon, any compound synthetically derived from them can rightly fall within the classification of organic chemistry.

The herbal flea collar calls itself "organic" to distinguish its herbal ingredients made from plants from toxic pesticides synthesized from petrochemicals.

In this context, "organic" is used to indicate the ingredients originally came from living organisms.

Foods and fibers labeled "organic" have been grown and processed in accordance with specific growing practices that use plant- and animal-based pest controls and soil amendments instead of synthetic petroleum-based pesticides or artificial fertilizers. It is this definition of organic that indicates a product is sustainable.

Organic agriculture is becoming more and more important as we seek to live in a sustainable world. Agribusiness agriculture has led to soil degradation and loss of fertile topsoil, excessive salt in fresh surface waters from soil leaching, loss of biological diversity, and contamination of ecosystems and poisoning from toxic pesticides.

More than 200 million tons of pesticides containing more than one hundred active ingredients are used annually on food croplands in California alone. In independent laboratory tests done in 1983 for the Natural Resources Defense Council (NRDC), 44 percent of the produce tested had detectable residues of nineteen different pesticides. Almost half had residues of up to four different pesticides.

Fiber crops are also heavily sprayed. Cotton is the most contaminated of all the natural fibers; seeds are treated with fungicides, herbicides are used repeatedly throughout the growing process, and heavy use of pesticides is common practice worldwide. Whatever pesticide residues might remain in cotton fabrics after processing seem to be harmless to health, but we must also consider the environmental destruction of our air, land, and water, and the illnesses these pesticides cause cotton growers and their families.

Two federal agencies hold responsibility for protecting us from pesticide residues in food: the Environmental Protection Agency (EPA) and the Food and Drug Administration (FDA). Under the Federal Insecticide, Fungicide, and Rodenticide Act (FIFRA), the EPA requires manufacturers of new pesticides to submit information about which crops they will be used on, their effectiveness and toxicity, and the nature and levels of residues. The EPA then sets a tolerance for each pesticide that establishes the amount allowed to remain on food crops.

The problem is that many pesticides now in use were granted tolerances before safety tests were required to determine whether the pesticide causes cancer, birth defects, genetic damage, or reproductive disorders. A study done by the National Academy of Sciences reports that 64 percent of pesticides now in use have not even been minimally tested for their toxic effects. Very little information is available on the potential long-term health effects, the possibility of synergistic effects resulting from exposure to more than one pesticide, and the range of individual sensitivities in the human population and other species. Of the nineteen pesticides detected in the NRDC survey to leave residues on produce, eight are suspected of causing cancer, five may induce birth defects, and the others could produce genetic mutations. But for some of the pesticides used on food, no information is publicly available to enable us to assess for ourselves the possible health hazards.

Under the Federal Food, Drug, and Cosmetic Act, the FDA tests samples of food crops for pesticide residues, but the tests commonly used by the FDA can only detect 107 of almost 300 pesticides that have established tolerances. As many as twenty pesticides suspected of causing cancer that are in the food supply are not even monitored by the FDA.

Further, some highly toxic pesticides are found in our food supply that are actually banned for use in the United States. Organochlorines such as DDT, 2,4,5-T, aldrin, dieldrin, heptachlor, chlordane, endrin, and BHC are illegal to use here but are still made by American manufacturers and shipped to other countries for use. They then return to us on imported foods—for example, coffee, chocolate, bananas, rice, tea, sugar, and tapioca. During the winter months, summer-season vegetables in supermarkets may be imported from Mexico, where pesticide use is not as tightly controlled as in the United States. These include zucchini, summer squash, yellow squash, garlic, string beans, okra, asparagus, bell peppers, cucumbers, eggplant, green peas, snow peas, and tomatoes. Even seasonally, if American crops fail or supplies aren't sufficient, green onions, radishes, parsley, corn, brussel sprouts, spinach, loose-leaf lettuce, watermelons, cantaloupes, honeydew melons, pineapples, and strawberries may also be imported.

ORGANIC, ORGANICALLY GROWN, BIODYNAMICALLY GROWN

Organically grown foods and fibers are those grown or raised without synthetic fertilizers, pesticides (including herbicides, insecticides, and fungicides), artificial ripening processes, growth stimulators and regulators such as hormones, or antibiotics and other drugs. In addition, organically grown foods must also be processed, packaged, transported, and stored without the use of chemicals such as fumigants, artificial additives, and preservatives, and without food irradiation. Organically grown fibers are specially processed to reduce or eliminate toxic substances in all phases of production.

While this is the general philosophy, there are many different methods of organic agriculture, which are specific to the site, the foods and fibers being grown or raised, the problems with pests, and the philosophy of the farmer. In general, though, organic growers maintain a program of reintroducing organic matter back into the soil by using compost, green manures, cover crops, animal manures, and fish meal. Minerals and other soil-building amendments such as seaweed and bone meal are often added to balance the soil to a healthy fertility. Crop rotation is used to manage weeds and plant disease, and natural predators, such as ladybugs and other beneficial insects and microorganisms, are introduced. Some organic growers plant unusual varieties to help maintain biological diversity. And most organic growers are small family farmers, concerned about developing regionally self-reliant food production systems, though larger growers are also starting to go organic.

Biodynamically grown food comes from farms designed to be self-contained sustainable ecosystems that bring together the complex interrelationships of plants, animals, and soil, with the warmth of the sun, the seasonal energies of the earth, and the rhythms of the cosmos. It is a special type of organically grown food that meets or exceeds all organic growing standards. Biodynamic concepts were introduced by philosopher Rudolph Steiner more than sixty years ago. Less known in this country than in Europe and other parts of the world, some biodynamic foods are sold in natural food stores and by mail.

Federal Regulations

The Organic Food Production Act of 1990 required the U.S. Department of Agriculture (USDA) to establish an organic certification program for producers and handlers of organically produced agricultural products. Since October 1993 it has been a violation of federal law to sell or label an agricultural product as organically produced, or to affix a label or provide other market information which implies directly or indirectly that a product is organically produced, unless the product is produced and handled in accordance with the requirements established by the program.

The act also requires all producers and handlers of organically produced agricultural products to be certified by a state or private certifier accredited by the USDA—so if it's not certified, it can't be called "organic," even though a farmer might use organic methods (if you come across food not labeled organic, but which describes organic methods, use your best judgment—there are many reasons why small growers choose not to be certified, even though their products meet, or even exceed, organic standards).

As is common in previous certification programs, the national certification requires a three-year period before the harvest of a certified crop during which the land is free from synthetic chemicals. Growers and processors who want to be certified have to submit a written plan of their organic practices. After harvest processors are not allowed to add any synthetic ingredients or ingredients that haven't been produced organically unless those ingredients are approved by the program or represent less than 5 percent of the finished product. They also cannot add sulfites to wine or foods, or nitrates and nitrites to processed meats or other foods. Fungicides, preservatives, and fumigants are forbidden in packaging, and even any water used has to meet federal standards for safety.

Livestock (including wild game, poultry, fish, and other forms of nonplant life) is also covered under this law. To be called "organic," only organically grown feed can be given—no growth promoters, hormones, antibiotics (except to treat illness), or other synthetic substances that are routinely used in this country. Dairy cattle have to be raised in this way for at least one year before their milk or products made from their milk are certified.

For processed foods, if 50 percent or more by weight of the ingredients are organic (excluding water and salt), the product as a whole can be labeled as organic. If less than 50 percent of the ingredients are organic, the word *organic* can be used on the ingredient list only next to those ingredients that are organic.

Each state is allowed to develop its own certification program, as long as it is at least as strict as the national program. Certifying organizations will have to be accredited by the Secretary of Agriculture.

If a grower or processor breaks the law, they cannot be recertified for five years and can be fined up to $10,000.

In the past organic crops have been certified by organizations of farmers, not the government, and it's been working very well. Most organic farmers have a sense of integrity about what they do and farm organically because they know it's the right thing to do and they want to do it right. It seems that the natural progress of certification is that many individual groups start certifying in their own area and then a larger organization—whether it is the government or a private group (see Forest Stewardship Council in Chapter 7)—wants to intervene to "make the standards consistent." While there certainly are "standard" principles that can be followed, there will always be a need to have growing practices and certifications reflect the particular characteristics of the place.

State Regulations

More than one third of the states have passed state legislation governing the use of the term "organic" for growing or labeling foods: California, Colorado, Connecticut, Idaho, Iowa, Maine, Massachusetts, Minnesota, Montana, Nebraska, New Hampshire, New York, North Dakota, Oregon, Rhode Island, South Dakota, Texas, Vermont, Washington, and Wisconsin. Since more states pass such legislation every year, it is likely that soon every state will regulate the use of this term. Because space does not permit details on the requirements of each of these laws, if your state has a law, get a copy and study it. Your local natural food store or any place that sells organically grown food should have a copy.

The state of California has one of the oldest laws regarding labeling of organic food, originally written in 1979 and revised in 1990. Many states have used the California standards over the years, and products all over the country are labeled "Grown and processed in accordance with section 26569.11 of the California Health and Safety Code."

Certification Organizations

Certification for organically grown products is way ahead of any other type of sustainable product. Certification organizations have been in place for more than a decade, and both local and national organizations exist across the country. These organizations have sprung "organically" from within the industry itself. Because those working with organic foods have a special care and dedication, they joined together to establish their own standards and certifications.

Certification that a product is organically grown is your guarantee that accepted organic methods were used. There are thirty-two certifying organizations for organic food in the United States and one—a branch of the Demeter Association in Germany—that certifies biodynamically grown food (national organizations are listed in the Directory of Organizations; to find out about state certifiers, ask your local natural food store).

There are two types of organic certifications—a full certification and a "pending" or "transitional" certification. Before granting a full certification there is generally a waiting period of three years between the time the grower switches to organic methods from chemical agriculture and when a full certification is granted. Certification logos usually indicate if the certification is pending.

While fully certified organically grown plants are somewhat healthier and contain fewer pesticide residues from the soil, buying organic products pending certification is just as important. Many farmers have some fully certified fields and some in transition, and as more and more farmers go organic, the number of fields with certification pending will far outnumber those that have full certification. Whether the field is fully certified or transitional, both use the same organic methods with the only difference being that the fully certified fields have not used pesticides and artificial fertilizers for at least three years. Transitional is just that, being in transition to organic, and needs to be supported as a midway step between chemical agriculture and organic agriculture. What is most important is that organic methods are used.

"Organic" is also used on fertilizers designed for use in organic gardens. I saw one misleading ad that said "CERTIFIED for use on ORGANICALLY GROWN lawns and crops." Upon closer examination, I saw that the brand name is "Certified®" and no other indication that the product was actually certified by an independent organization. So be sure that if a product says "certified" it also says who it's certified by, or that the company will furnish that information on request. As more and more organic foods and fibers are grown and made into products by large manufacturers, there will be times when raw materials from many different growers and certifiers used to make a line of clothing, for example, will be too many to list on the label. In those cases, the manufacturer should be able to supply a list of certifiers on request.

Scientific Certification Systems has certified food under its NutriClean program since 1984, but they do not certify foods to be organically grown. Instead, their certification is for a different standard they call "clean food"—food certified to contain no detectable pesticide residues.

The purpose of the NutriClean system is to screen produce for pesticides that have the greatest health concern to the public, measuring final crop performance against a strict "no detected residue" standard. Their comprehensive audit and review process includes complete disclosure of the pesticides used by the grower, on-site inspections, field sampling of products, and extensive laboratory tests for each and every pesticide used.

NutriClean standards for pesticide use are stricter than government or some organic standards because they do not allow residues of either synthetic pesticides or the natural pesticides sometimes used in organic agriculture. The certification is limited, however, because it addresses pesticide residues only and not the fertility of the soil or the overall sustainability of the farm.

Organically Grown and Biodynamically Grown Products

After years of having a wilted image, organically grown **FOOD** is now becoming big business. In 1989 the National Academy of Sciences found that organic farming can produce just as much food and profit as conventional agriculture, but it wasn't until Meryl Streep alerted the nation to Alar in apples that even supermarkets started carrying organically grown food. Between 1986 and 1991 the number of new organically grown food and beverage products introduced increased more than 400 percent, but organically grown food still only represents about 1 percent of all the food grown in this country.

In addition to organically grown food, **WINE** is now available made from organically grown grapes. Some is imported from France, Germany, and Italy, but quite a lot is made in California.

And organic methods have recently spread into cotton growing, too (see **TEXTILES**).

WILDCRAFTED

Another term used to identify pesticide-free foods is wildcrafted. Wildcrafted plants are, as the term suggests, gathered from their natural, wild habitat, often from pristine areas. Sea vegetables, wild rice, some wild fruits, and many herbs are wildcrafted.

When wildcrafting is done sustainably with proper respect, generally only the branches or flowers from plants are taken and the living plant is left, or if it is necessary to take the whole plant, seeds of the plant are placed in the empty hole from which the plant was taken. Care is taken to only remove a few plants, flowers, or branches, so plenty remain to continue the supply.

Wildcrafted plants are regulated by The Organic Food Production Act of 1990. Harvesters must designate the area they are harvesting and provide a three-year history of the area that shows no prohibited substances have been applied there. A plan for harvesting must show that the harvest will sustain the wild crop. No prohibited substances can be added by processors.

Wildcrafted Products

Wildcrafted plants are used mainly in **FOOD** products and herbal cosmetic preparations (see **BEAUTY AND HYGIENE**).

CHAPTER 6

Reduced
Reusable
Refillable
Reclaimed
Recycled
Recyclable

Recycling is one of the basic precepts of sustainable living. Nature recycles everything 100 percent. Every blade of grass, butterfly, fish, leaf, bird, all degrade back into the soil and then those broken-down nutrients are used to make new life forms. In a completely sustainable world, all the products we use would either be biodegradable and go back into the earth (see Chapter 11) or they would be recycled into new products. Everything would be useful—there would be no waste.

When we think of recycling, our first thought often is of garbage—recycling is collecting our bottles, cans, and papers to be taken to the recycling center instead of a landfill. But from a sustainability point of view, recycling is also a gold mine of raw materials from which new products can be fashioned. Instead of only thinking of recycling when throwing away a product, we need to think as well of recycling when evaluating the materials from which a product is made.

There's nothing new about recycling—in some form or another it has been around for centuries. It used to be common practice for scrap merchants and scavengers to collect what others no longer wanted and either repair the item or turn it into something else that could serve a useful purpose. The scrap recycling industry estimates that people were reusing metal scrap as far back as five thousand years ago and the reuse of wastepaper and textiles dates back more than three thousand years.

During times of crisis, when raw materials and products are in short supply, reuse and recycling always come back. While I was born and raised in the disposable age, I remember my grandparents and great aunts and uncles all saved and reused everything, a habit learned during the Depression and World War II.

Now that we are faced with the need to conserve resources and limited landfill space, recycling, I believe, will become as normal a part of our lives as taking out the garbage. In the last few years, the collection of recyclables and the manufacture of recycled products has become a multimillion-dollar industry. We are beginning to see that recycling is the most effective and economical way to handle the waste we produce in this country.

A very large portion of our waste is recyclable or otherwise reusable. If we recycled all of our paper and paperboard (40 percent), all of our metals (9 percent), all of our plastic (8 percent), and all of our glass (7 percent), we would reduce our total garbage by 64 percent! If, in addition, we reclaimed our wood (4 percent), rubber (3 percent), and textiles (2 percent) for other uses, we would be down to only 27 percent of our current garbage. By composting (see Chapter 11) yard wastes (18 percent) and food wastes (7 percent), we're left with only 2 percent of our current solid waste to place in a landfill or otherwise dispose of.

U.S. GARBAGE

Municipal Solid Waste (by weight)		*Where It Goes*
40% paper and paperboard		73% landfills
18 yard wastes	→	14 incinerated
		13 recycled
9 metals		
8 plastics		
7 glass		
7 food wastes		
4 wood		
3 rubber, leather		
2 textiles		
3 miscellaneous		

Source: Data from the EPA.

Of course, this is idealistic, but it's a sustainability goal worth pursuing. Though the national recycling average is only 13 percent, some individual states have higher recycling rates: Washington 28 percent, New Jersey 25 percent, Oregon 25 percent, Vermont 18 percent, Illinois 18 percent, and Maine 17 percent. It can be done.

While we are a long way from the ideal, in the years to come more and more products will be made from recycled materials. In addition to preserving resources and saving the rain forest (many U.S. paper mills import tropical wood pulp to make paper), recycling also reduces litter, conserves energy and water, saves on packaging costs, and lowers community disposal bills. Recycling also creates jobs, as it is more labor-intensive than dumping the same material in a landfill.

Despite the clear advantages, in the past recycling has not been widely adopted, primarily because the cost of operating ongoing recycling programs has often exceeded the market of manufacturers needing recycled materials. But our perceptions about recycling are rapidly changing as we face the enormous problem of what to do with all the garbage we create in our disposable society. A new awareness of recycling is emerging as our trash is becoming recognized as a valuable source of raw materials.

Recycled and recyclable products have an almost universally recognized symbol—the three "chasing arrows" (see sidebar on page 104). It was originally developed in the 1970s for use on recycled paper, but it is now widely used in various forms on products of all kinds and has universal recognition as the symbol for the recycling of anything and everything. New York has trademarked its own version of the symbol and regulates its use. Some other states also regulate use of the symbol. But don't make any assumptions about what this symbol means—I've seen it used on products to indicate both that the product is made from recycled material and to indicate that the product can be recycled (two very different things). Without descriptive words explaining the meaning of the symbol, it is practically worthless beyond a general indication that the product somehow can be part of the recycling loop.

The term *recycled* has common meanings and legal meanings. In general usage, we say we are "recycling" something whether we take it to a recycling center or simply use the product again.

The most important point about recycling is that it reduces the amount of raw materials used to make products, and the amount of waste that needs to be landfilled. But there are other ways we can accomplish these same goals, too.

Most sustainable is simply to reduce. We can all use less, whether that means fewer products or products that use less material in the package or the product itself. Many products and packages today are being redesigned to be more resource-efficient.

Next most sustainable is to reuse. Many products can be used multiple times or for multiple functions, and materials can be reclaimed to make new products without going through additional manufacturing.

And we can recycle our products and complete the circle by buying products that contain recycled material.

REDUCED

We can reduce our use of resources and the amount of garbage we create both by simply buying less and by using products that make efficient use of resources. Regardless of what material is used to make the product, the fact that less material is used makes a product more sustainable. No product or package should use more resources than is necessary to do the job.

Source reduction claims are used to indicate that some reduction has been made in the amount of material used (over what was used previously) to make a product or package. Many products and packages are being redesigned to reduce the amount of material used. Today's aluminum can, for example, uses approximately one third less metal than the average can did in 1972, and major brands of laundry detergent and fabric softener are now sold in concentrated form in a smaller package.

Federal Guidelines

There are no federal regulations for source reduction claims; however there are guidelines for product labeling.

The FTC guidelines state "Source reduction claims should be qualified to the extent necessary to avoid consumer deception about the amount of the source reduction and about the basis for any comparison asserted."

A clearly stated claim would say something like "now 10 percent less material than our previous package." In this case the advertiser should be able to substantiate that because the package is made from 10 percent less material, disposal of the current package contributes 10 percent less waste by weight or volume to the solid waste stream when compared with the immediately preceding version of the package.

On the other hand, if an advertiser claims that disposal of its product generates "10 percent less waste," the claim is ambiguous and deceptive. Depending on the context, it could be a comparison either to the immediately preceding product or to a competitor's product. Without clarification the claim is meaningless, yet gives a false impression that the product actually has an environmental benefit.

State Regulations

Source reduction is defined by law only in the state of Indiana, where it means a reduction in the quantity or concentration or components of solid waste through actions affecting the source of solid waste before the product is generated. Minimizing, downsizing, or making a package more lightweight all qualify as source reduction, as does eliminating certain materials as packaging constituents, or concentrating the product to result in the need for less packaging.

Reduced and Reducing Products

Source reduction can be accomplished by buying products that use fewer resources. Buying in bulk or concentrated form or in family-sized containers to reduce packaging, avoiding single-use disposable products, and buying durable goods that can be repaired are all ways to reduce. Many products at the supermarket now come in concentrated form (particularly laundry products), and discount warehouses are full of institutional-sized containers of food and home products.

Another way to practice source reduction is to simply use fewer products. Don't buy ten different cleaning products when you can do most of your household cleaning with baking soda, vinegar, and soap. Begin to pay attention to everything you buy and use. Stores are filled with products they want consumers to buy on impulse—watch yourself and see if you are buying things you really want and need.

If you use certain items only occasionally—garden tools, sewing machine, party dishes—reduce the need to own your own by renting them or sharing with family, friends, and neighbors.

REUSABLE AND REFILLABLE

Many people say that an item has been "recycled" if they use it more than once—recycling an old peanut butter jar to store leftovers, for example, or recycling the back of an envelope for telephone messages.

While this is fine as a common use of the word, more correctly the term is *reused.* A reusable product is one that is designed to be durable and used over and over again, instead of being disposable and thrown away after one use (though some disposable products can be used more than once—you can get three or four good shaves out of a disposable razor and bake several pies in a disposable pie plate).

The importance of designing and purchasing products and packages that are reusable became abundantly clear to me one day as I was reading *Beyond the Limits: Confronting Global Collapse, Envisioning a Sustainable Future.* In it I read that as a rule of thumb, you can assume that for every ton of consumer garbage there are five tons of waste at the manufacturing stage and twenty tons of waste at the site of initial resource extraction (mining, pumping, logging, farming). That's twenty-five pounds of garbage for every pound of garbage we throw away! When I read this, I had a vision of twenty-five garbage cans on my patio instead of one, and this really brought home to me how much waste I actually produce. Not counting our municipal waste (what we throw away at home), the U.S. economy in 1985 (it's probably more now) generated:

628 million metric tons industrial waste (including 265 million tons hazardous waste),

72 million tons energy waste,

1,400 million tons agricultural waste,

1,300 million tons mining waste (excluding coal),

98 million tons demolition waste,

8.4 million tons sewage sludge.

Because it takes so many resources to make a product and the manufacture produces so much waste, it's important that every product made be durable and reusable. In this context, disposable products appear even more wasteful.

Increasing product lifetime through better design, repair of broken or worn items, and continuing to reuse a product for its entire life span is more effective than recycling, because it doesn't require crushing, grinding, melting, purifying, and refabricating recycled materials. Simply doubling the lifetime of any product will halve the energy consumption, the waste and pollution, and the ultimate depletion of all the materials used to make it.

Refillable is a specific kind of reusable package that can be refilled with a product over and over again. The refillable container that immediately comes to mind is, of course, the old glass milk bottle. For many years milk was not available in glass bottles, but glass milk bottles are now enjoying a renaissance in many areas. We have a local creamery that is using old glass milk bottles that have been collected from old dairies. It's a delight to discover the old dairy names still printed or molded into some of the bottles. Glass to be used in refillable containers is made about 50 percent heavier and can be reused up to thirty times.

One of the problems preventing greater use of refillable glass containers is that the size and shape and texture of the bottle or jar is often used as a marketing tool to help identify the product. Environmentally speaking, it would be a lot better if those making and using glass packaging would come up with some standard shapes and sizes that are economical for all to use, and establish the product identity with a creative, removable label. In Europe I saw a milk bottle with a label that fit around the top like a collar, which then slipped right off and allowed the bottle to be reused. With a little ingenuity, our package designers could greatly increase the amount of glass packaging that is available for reuse.

Federal Guidelines

I could not find any federal guidelines or regulations regarding the term *reusable*; however, the Federal Trade Commission does give guidelines for *refillable*.

They say "an unqualified refillable claim should not be asserted unless a system is provided for: (1) the collection and return of the package for refill; or (2) the later refill of the package by consumers with product subsequently sold in another package. A package should not be marketed with an unqualified refillable claim, if it is up to the consumer to find a new way to refill the package."

The FTC only wants a package to say it is refillable if it is part of a regular program (such as milk or soda bottles that are regularly collected and refilled), or if the original container can be refilled with the same product purchased in bulk or in a larger-sized container, or in concentrated form. Even though we could refill (or reuse) a container at home by our own choice, this home refillability does not fall within their definition.

They give a couple of examples. If a container is labeled "refillable X times," then the manufacturer must have the capability to refill returned containers and be able to show that the container will withstand being refilled at least X times. If the manufacturer has established no collection program, however, the claim is deceptive because there is no means for actual collection and return of the container to the manufacturer for refill.

Another example is a bottle of fabric softener that states it is in a "handy refillable container." The manufacturer also sells a large-sized container that indicates that the consumer is expected to use it to refill the smaller container. If the manufacturer sells the large-sized container in the same market areas where it sells the small container, the claim is not deceptive because there is a means for consumers to refill the smaller container from larger containers of the same product.

State Regulations

A handful of states have established regulations defining a product or package that is reusable or refillable.

In Indiana a reusable product or container is one which may be reused in its original form by consumers or others. Reusable/refillable packaging is that which is designed to be reused or refilled *for its original purpose* in an existing program, such as milk bottles.

In New York and Rhode Island the original package or product must be returned for refilling or reuse a minimum of five times, and this must be demonstrated by an annual accounting in a program established by a manufacturer, distributor, or retailer. The model legislation created by the Northeast Recycling Council agrees.

A proposed standard in Maine specifies that the package or material must be reused or refilled for its original purpose. So a milk bottle would be reusable if refilled with milk and sold again; however, those French jam jars that many people use for drinking glasses could not be called reusable.

In New Hampshire the original package or material must be used or refilled for its original purpose a minimum of five times.

In California a "post-filled" container is one "previously filled with a beverage or food."

In most states the chasing arrows recycled symbol can also be used to indicate that the product is reusable, adding to the confusion between reusable and recyclable products and packages.

Reusable Products

There are many, many reusable products on the market, even though they are not always labeled as such because they are not part of an established program. Observe your shopping habits and see how many disposable products you buy. Here are some suggestions for replacing disposable products with durable ones:

DISPOSABLE	DURABLE/REUSABLE
Paper plates and cups	China, pottery, or glass dishes and glasses
Paper napkins	Cloth napkins
Plastic flatware	Stainless steel or silver flatware
Paper towels	Cloth towels and rags
Paper or plastic grocery bags	Cloth tote bags
Plastic food wrap	Reuse food jars, bottles, and tubs

DISPOSABLE	DURABLE/REUSABLE
Aluminum foil	Baking pans
Disposable diapers	Cloth diapers
Plastic razors	Refillable or electric razors
Plastic pens	Refillable fountain or ballpoint pens
Women's sanitary products	Cloth menstrual pads
Household batteries	Rechargeable batteries
Tissues	Cloth handkerchiefs
Paper coffee filters	Cloth or metal mesh filters

Seek out and buy good quality products that will last a long time. When you buy less expensive items, you may initially save money, but in the long run you may have to replace the less expensive item two or three times in the same time period the better quality item would last. And each time the cheaper product is replaced, a new one has to be manufactured using more resources and creating more waste. Look for products with lifetime guarantees—if a company promises to repair a product for your lifetime, chances are it will last longer before it needs fixing.

Another way products can be given a second (or third or fourth life) is by being repaired, rebuilt, restored, or reconditioned and then reused: Typewriter and computer printer ribbon cartridges can be reinked or refilled, tires can be retreaded, cars can get new engines, etc. For many products, individual parts can be reused to repair items—auto dismantlers particularly salvage the useful parts of crashed cars for resale, rather than simply recycle the whole car as scrap metal.

Some products can be remade through what is known as "remanufacturing," an industrial process in which products are reassembled using old restored parts and a few new parts to produce a unit that is of the same quality as, if not superior to, the old. Currently automotive parts, industrial equipment, office machinery, and appliances form the largest market of remanufactured products.

You can also reuse products by purchasing used (or "preowned" as the sign on my local Mercedes Benz dealer says) or refurbished items. Antique stores, flea markets, garage sales, storage auctions, classified advertisements, and community bulletin boards are filled with products that may be of interest to you. If you have no use for an item that is still serviceable, take it to a flea market, give to a thrift store (and take a tax deduction), hold a garage sale, or give it to a community service group. Sell or give away items you no longer use to allow them to be of use to someone else. Periodically clean out your garage, cabinets, and closets, and put things you no longer use back into circulation.

RECLAIMED

Reclaimed means a material is being reused to make a product, but in a different form. Halfway between reused and recycled, a reclaimed material is changed from its previous form, but not entirely reprocessed. A classic example of a reclaimed material is fabric scraps used to make patchwork quilts, or the calico dresses made out of flour sacks.

As far as I know, there are no regulations or guidelines regarding the use of the word *reclaimed* on product labels.

Reclaimed Products

Some reclaimed products in today's marketplace include birdhouses made from wood reclaimed from old barns, stationery made from reclaimed out-of-date maps, and **BARBECUBE** briquets made from fruit tree orchard prunings.

RECYCLED

A recycled product is a new product made from materials that would otherwise have been waste, broken down (melted or pulped) into a basic substance from which a new product can be formed. Even though the word *recycled* is commonly used to mean reusing something in any way, legal definitions require the old product to be broken down into its basic material and a new product formed before it can be called "recycled."

Recycled products are the other half of the recycling cycle. In order for product recycling to be successful, the recyclable items must be 1) collected, 2) delivered to and used by a manufacturer to make a new product, and 3) the new product containing the recycled material must be purchased by a consumer. A discard is a discard until somebody figures out how to give it a second life, and a market exists to give that second life economic value. So recycling is more than taking your newspapers and soda cans to the recycling center—it's buying products made with recycled material, too.

There are three general levels of recycling.

The first (appropriately called "primary recycling") is the reprocessing or remanufacturing of discarded materials into the same product which can then be recycled again, such as a glass container into a glass container or a steel product to a steel product.

The second level ("secondary recycling") is the reprocessing or remanufacturing of discarded materials into a different, but similar, product which is technically recyclable—old corrugated cardboard boxes into cereal boxes, for example.

The third level ("tertiary recycling") is the reprocessing or remanufacturing of discarded materials into a product which is not likely to be recycled, such as recycling mixed office paper into bathroom tissue.

There is much controversy among manufacturers, distributors, regulators, and consumers over what constitutes a "real" recycled product. There are two kinds of recycled material that can go into making a recycled product. "Post-industrial waste" or "pre-consumer waste" is that waste generated by industrial manufacturing processes that would have otherwise gone to a landfill (not included are wastes that have customarily been "put back in the pot" in the factory). "Post-consumer waste" or "PCW" is the bottles, cans, plastic milk jugs, and newspapers we collect and recycle after we are done using them.

There are those who believe that only products made from post-consumer waste should be called "recycled." Considering that for every pound of consumer waste there is twenty-five pounds of manufacturing waste, I think that pre-consumer waste has its place in products called "recycled," too. Products just need to be labeled correctly, so we know what we're buying.

Federal Guidelines

The Federal Trade Commission and the Environmental Protection Agency have guidelines for defining recycled products.

Federal Trade Commission

The FTC gives guidelines for claims regarding recycled content. They say "A recycled content claim may be made only for materials that have been recovered or otherwise diverted from the solid waste stream, either during the manufacturing process (pre-consumer), or after consumer use (post-consumer) [material that is routinely collected and "put back into the pot" cannot be included]. To the extent the source of recycled content includes pre-consumer material, the manufacturer or advertiser must have substantiation for concluding that the pre-consumer material would otherwise have entered the solid waste stream. In asserting a recycled content claim, distinctions may be made between pre-consumer and post-consumer materials. Where such distinctions are asserted, any express or implied claim about the specific pre-consumer or post-consumer content of product or package must be substantiated."

The guidelines go on to say that unqualified claims may be made "only if the entire product or package, excluding minor, incidental components, is made from recycled material. For products or packages that are only partially made of recycled material, a recycled claim should be adequately qualified to avoid consumer deception about the amount, by weight, of recycled content in the finished product or package."

Here are a couple of their examples which show acceptable claims that would not be considered misleading.

A greeting card is composed of 30 percent by weight of paper collected from consumers after use of a paper product, and 20 percent by weight of paper that was generated after completion of the paper-making process, diverted from the solid waste stream, and otherwise not normally reused in the original manufacturing process. The marketer of the card may claim either that the product "contains 50 percent total recycled material," or may identify the specific preconsumer and/or post-consumer content by stating, for example, that the product "contains 50 percent total recycled material, 30 percent of which is post-consumer material."

A package with 20 percent recycled content by weight is labeled as containing "20 percent recycled paper." This would mean that some of the recycled content was composed of material collected from consumers after use of the original product, and the rest was composed of overrun newspaper stock never sold to customers or some other kind of pre-consumer material.

Many products that contain some post-consumer material do not specify the percentage because it varies from batch to batch due to availability. A common way recycled products are labeled is "contain —% recycled material with a minimum —% post-consumer material." This allows manufacturers to vary the amount of post-consumer material used, increasing it as more becomes available.

Environmental Protection Agency

Under Section 6000 of the Solid Waste Disposal Act, as amended by the Resource Conservation and Recovery Act of 1976 (RCRA), federal agencies purchasing certain designated items must endeavor to ensure that these items are composed of the highest percentage of recovered materials that is practical and possible at the time. Section 6002 of RCRA requires the EPA to designate such items and to prepare guidelines to help procuring agencies comply with the act's requirements.

In 1988 the Environmental Protection Agency issued the EPA Guidelines for Procurement of Products that Contain Recycled Materials to aid government procurement officers in the purchase of recycled products.

The EPA Guidelines define *recovered materials* as being "post-consumer materials" and "manufacturing, forest residues, and other wastes." Post-consumer materials are "items which have passed through their end-usage as a consumer item." Manufacturing, forest residues, and other wastes are pre-consumer wastes, including "manufacturing wastes . . . forest residues from manufacturing or woodcutting processes . . . waste rope from cordage manufacture and textile and mill waste and cuttings used in production of cotton fiber papers."

According to this EPA definition, sawdust could be used in making paper and be counted as recycled content. There has been much controversy within the recycled paper industry about this definition—while it's fine to use sawdust to make paper, sawdust isn't recycled *paper* by any stretch of the imagination, and I'm surprised the FTC doesn't consider this deceptive.

In 1992 recycled paper standards and labeling guidelines were issued by the Recycling Advisory Council (RAC), a U.S. Environmental Protection Agency–funded offshoot of the National Recycling Coalition (NRC) that includes members from the manufacturing, government, and public interest sectors.

To encourage nationally uniform, consistent labeling as a means to promote recycling, RAC recommends that manufacturers using recycled materials in their paper products or packages follow some general labeling guidelines:

- Paper products and packages should bear labels disclosing the percentage of both total recycled fiber and post-consumer fiber
- Products and packages should use the chasing arrows recycling symbol only if the item meets both the recommended total recycled content and post-consumer recycled content standards
- Manufacturers making recycled content claims should be able to prove their claims
- Labels should indicate whether a recycled content claim refers to the product, to the package, or both
- Labels should not be misleading or deceptive.

These guidelines, I believe, could be agreed upon by most people in the industry as being reasonable, responsible, and informative.

State Regulations

In some states there are legal definitions for the word *recycled*, for determining recycled content, and for other terms associated with labeling recycled content.

In Indiana recycling includes not only the diversion of materials that would become or otherwise remain waste from the solid waste stream, but also the beneficial use of such materials in a new product. Once these materials are collected, separated, and processed, they must be used as raw materials or feedstocks in lieu of or in addition to virgin materials in the manufacture of goods sold or distributed in commerce. According to Indiana law, recycling does not include burning municipal solid waste in incinerators for energy recovery.

In California and Indiana a recycled product must contain at least 10 percent post-consumer material by weight.

New Hampshire (and as proposed in Maine) requires that the label indicate the percentage by weight of pre- and post-consumer composition in order to use the term *recycled* or the recycled symbol.

In Minnesota a person selling a product or package may indicate that the item contains recycled material only if the label states the minimum percentage of post-consumer material in the product or package: by weight for a finished nonpaper product, and by fiber content for a finished paper product or package.

In Rhode Island the recycled content is that portion of material or a package's weight that is composed of pre-consumer and post-consumer materials.

Almost all the states that define pre-consumer material agree with the Federal Trade Commission definition. In Maine, New Hampshire, New York, Rhode Island, and the model legislation written by the Northeast Recycling Council, pre-consumer material is any material generated during any step in the production of an end-product that has not reached a business or consumer for an intended end use and has been recovered or diverted from the waste stream. Waste materials or by-products that can be or have been normally reused within the same plant or another plant of the same parent company are not included, unless the material has been printed or dyed in the manufacturing process. In the paper industry pre-consumer materials do not include wet or dry mill broke, rejected unused stock, obsolete inventory, butt rolls, or other paper waste generated by the paper product mills. Also not included is waste generated by converting operations that are used by the same parent company whether for the same or different product.

The definition for post-consumer material or post-consumer waste also has consensus. California, Indiana, Maine (proposed), Minnesota, New Hampshire, New York, Rhode Island, and the Northeast Recycling Council agree that post-consumer waste is material generated by its final consumer, which has served its intended end use, and has been diverted from the solid waste stream through a recycling program.

Certification Organizations

Both Scientific Certification Systems and Green Seal have issued standards regarding recycled content. The standard from SCS can be applied to all products with recycled content; the Green Seal standard applies only to paper (see **PAPER**).

The SCS certification for recycled content may be awarded to consumer products, product packages, or the intermediate materials from which such products or packages are made. The percentage of recycled post-consumer waste used must be at least 10 percent by weight or greater if required by law in the areas where the product or package is sold. In addition, the total recycled content (pre- and post-consumer material) must fall within a "state-of-the-art" range (equal to or exceeding 80 percent of the highest percentage of recycled material that can be technically and operationally incorporated into a specific product on a significant scale). The product must also be recyclable, unless it cannot be made into anything but a nonrecyclable form (such as facial tissues).

An Exceptional Recycled Content Label

As manufacturers are becoming more aware that consumers are interested in knowing the environmental attributes of products, we are getting more and more information. One new company with excellent product disclosure is Deja, Incorporated, makers of **Eco Sneaks** and **Envirolite** shoes for men and women.

This new company is dedicated to sustainable development using high-tech recycling technology. Their marketing materials completely describe the wide range of recycled materials used to make their shoes:

- Upper/outer fabric and inner lining material in the ECO SNEAKS made from 100% recycled cotton canvas from ECO FIBRE, a new high-quality yarn spun from recovered and recycled natural and synthetic fibers.
- Molded upper rubber parts constructed from 60% thermal plastic rubber, 30% of which is pre-consumer recycled polypropylene from the trim waste during the manufacturing of diapers.
- Developed by TEXON, the lasting board is made from 100% recycled fibers manufactured from reject coffee filters and file folders.
- In the ECO SNEAKS, the interfacing material between the outer fabric and the inner lining employs a non-woven polyester material that is 80% post-consumer recycled polyester soda pop bottles.
- Foam used in the upper collar area and tongue is 100% recycled materials from seat cushions and chairs.
- The heel counter stiffener is made from trim waste from the resin materials used by other shoe makers. DEJA's supplier has agreed to buy back and reprocess from waste from other shoe manufacturers on behalf of the company.
- The neoprene foam cushioning attached to the lasting board is 100% pre-consumer recycled trim waste from wetsuit and gasket manufacturers.
- The shank board, which provides the shoe's stability, is made from 100% recycled grocery bags and corrugated cardboard.
- For the outsoles, DEJA is using a new process binding 50% pre-consumer recycled rubber, destined for incineration or the landfill, with 50% virgin rubber. Research is under way to create a material with an even higher percentage of recycled rubber.
- Midsoles are 40% pre-consumer recycled ethyl vinyl acetate foam made from other shoemaker's scrap, which may have otherwise gone to a landfill.
- Laces are 100% cotton.
- The unique DEJA shoe box is 100% recycled corrugated cardboard with water-based inks used in the printing. To encourage customers not to throw it away after purchasing the shoes, the decorative shoe box is ingeniously designed to be used as a

gift box. The box is constructed so that it can be easily folded inside out, revealing beautiful etchings of rare and endangered animals.

- When the shoes eventually wear out, customers can send them back to DEJA to be recycled again. "We'll encourage customers to recycle their old shoes by sending them back to us. By doing this, we can further control the entire life cycle of our products."

Wouldn't it be wonderful if every product had such a label?

Recycled Products

One of the most important changes we need to make in our buying habits is to purchase products made from recycled materials. Since Earth Day 1990 there has been a boom in both recycling programs and citizens who are actively recycling. But most of the attention has been paid to collecting recyclables rather than having a balanced campaign that also promotes the purchase of products containing recycled material. The result is that now there is a surplus of recycled material that needs a market.

At least ten states and six hundred communities currently have mandatory recycling laws, and these numbers can only grow. Recycling may soon become mandatory across the nation. But where will all this recyclable material go? Already prices are dropping for recycled materials, jeopardizing existing programs. We have to buy more recycled products. It's up to each of us as consumers to keep the recycling loop flowing by putting recyclables in and taking recycled products out. If you're not buying recycled products, you're not recycling.

Look for and purchase products made from the following recycled materials.

Aluminum

Of all the various metal products that can be recycled, the most widely recycled by far are aluminum cans.

Most aluminum goes back into a perpetual cycle of making soft drink and beer cans. It is very easy to recycle because there are no labels, caps, or tops that must be removed. According to the Aluminum Recycling Association, in 1991 we reached a national recycling rate for aluminum cans of 62 percent, although California and Texas are nearer to 70 percent. Aluminum cans can be remelted and back on the supermarket shelves in as little as six weeks. The aluminum industry claims that cans average more than 40 percent post-consumer content, though this fact is not generally noted on the label.

Aluminum is also beginning to be used to build recyclable automobiles (see AUTOMOBILES).

Cloth

Recycled fabrics are made from pre- and post-consumer wastes, which are collected, unraveled, then respun into yarn for weaving or knitting (see TEXTILES).

Glass

The National Soft Drink Association estimates that more than 5 billion glass bottles and jars are recycled each year. Glass is 100 percent recyclable—bottles and jars can be melted down and turned into new containers without adding any additional ingredients. One ton of glass returned to a recycling plant can produce one ton of new glass, and save 1.2 tons of raw materials. Because glass never deteriorates, it can be recycled endlessly.

Industry wide, approximately 30 percent of manufactured glass bottles and jars are made from post-consumer recycled glass, which means that the average glass container you buy in the store has about that much post-consumer recycled content, whether or not the recycled content appears on the label. Manufacturers could use a higher percentage; however their supplies of recycled glass cullet are inconsistent. Because the furnaces used to melt glass must be set to a different temperature depending on the amount of recycled cullet used, manufacturers use a consistent average of 25 to 30 percent, even though at times they may have up to 70 or 80 percent available.

The glass industry is very much in favor of using recycled cullet as there are cost benefits to it. We just need to improve our recycling rate for bottles and jars, and eventually true recycling can be achieved and all glass will be made out of recycled material.

Some broken glass of mixed colors that cannot be recycled back into glass is used in the production of fiberglass and as a substitute for stone in glasphalt.

A California law requires all glass containers made or sold in the state to contain specified minimum percentages of post-filled glass. Currently the law requires 25 percent, but it goes up to 35 percent by January 1, 1996, 45 percent by January 1, 1999, 55 percent by January 1, 2002, and 65 percent by January 1, 2005.

Motor Oil

Used motor oil is now being collected and recycled in many communities. Recycling keeps hundreds of millions of gallons of used motor oil from polluting our land and water—used motor oil contains heavy metals like lead and cadmium. One gallon of used motor oil can contaminate up to a million gallons of fresh water. When motor oil reaches fresh water, even the thinnest film on the water surface blocks sunlight, prevents the replenishment of oxygen, and impairs photosynthesis.

We only recycle about 30 percent of the motor oil we use. The rest is discarded—often poured down a storm drain, put in a trash can to end up in a landfill, or poured directly on the ground. It has been estimated that the amount of used motor oil dumped in the United States annually is ten to twenty times the amount spilled from the *Exxon Valdez* tanker in Alaska in 1989.

Used motor oil is recycled simply by heating it to remove impurities and then it is ready for reuse. Two and a half quarts of recycled motor oil can be made from one gallon of used oil; it takes forty-two gallons of virgin crude oil and a lot more energy to make that same two and a half quarts (see **Motor Oil**).

Paint

While most brands of paint you'll find on the shelf have been formulated from virgin materials, there are paints available in some communities that are "recycled" remixtures of paints that have been turned in during household hazardous waste collection days.

Paper

The recycled product we most need to support with our purchases is recycled paper. Paper accounts for 37 percent of the total waste produced in the United States. Approximately 24 percent is now recycled (during World War II we recycled 43 percent). Instead of piling up in a landfill, paper can be recycled seven to ten times before the fibers become too soft to hold together.

Americans use 50 million tons of paper each year, consuming more than 850 million trees. Recycling half the paper used throughout the world today would free 20 million acres of forest land from paper production.

Throughout history paper has mostly been made from recycled materials. Ancient Chinese historical documents record that the source of fiber for Ts'ai Lun's paper included discarded cloth and rags. The first recorded use of wastepaper for making new paper dates from A.D. 1031 in Japan, where recycled paper was held in very high regard.

After Gutenberg's invention of the printing press in Europe, there was a great demand for paper. At first paper was made exclusively from old linen rags, but by the mid 1700s rags made from imported cotton were also used. When the first paper mill in America was constructed in 1690, old linen rags were also the main source of fiber, although as cotton plantations developed in the South, cotton rags were also used. Near my home we enjoy many acres of redwood forest where a paper mill once stood—the paper was made from rags collected locally instead of cutting down the trees.

With the beginning of industrialization and the settling of America, there

PAPER RECYCLING SYMBOLS

The three chasing arrows were first developed in the early 1970s by the American Paper Institute as a means to promote recycling by creating consumer demand for recycled products. It was placed in the public domain by the paper industry to encourage broad recognition and usage.

In response to growing public interest in recycled papers and the increasing availability of paper products containing recycled material, the Institute (now called the American Forest and Paper Association) developed new guidelines for using this symbol in 1993.

The new symbols are designed to identify recycled fiber content paper and paperboard products and to provide special recognition for products manufactured from 100 percent recycled fiber. It is recommended that the symbols be accompanied by an information legend that explains the symbol.

The 100 Percent Recycled Product Symbol is reserved for use only on those products that are manufactured with 100 percent recycled fiber.

The Recycled Content Product Symbol may be used with paper and paperboard products which contain a percentage of recycled paper fiber. The terms *total recycled fiber, total recycled paper,* and *total recovered fiber* may be used interchangeably in the legend.

Be cognizant, though, that many paper products may be mislabeled from the past. Don't make assumptions as to the meanings of the chasing arrows if they do not have accompanying information.

was a tremendous increase in the use of paper, and a shortage of rags. By 1880 many American newspapers were printed on paper made from wood pulp, and as mechanization made the process cheaper and cheaper, paper made from wood became the standard.

With the shortages of World War I, manufacturers relied on the public for old rags and wastepaper to make paper. Used paper became a valuable and salable commodity as thousands of tons of old books, newspapers, and business papers were recycled by the mills. Recycling surged again during the shortages of World War II. But throughout the century the small mills continuing to make recycled paper have had difficulty competing with the economies of scale of the large paper corporations, and for a while, recycled paper was all but impossible to purchase. With current and future concern about resource use and landfill space, recycled paper is enjoying a renaissance that could establish it firmly in the marketplace. In 1993, for the first time, more paper was recovered for recycling or reuse than was sent to landfills.

While recycled papers are much more common now than in the recent past, they still aren't a commonplace choice. The market has been in a catch-22—consumers want recycled paper to be more widely available before they commit to using it, and paper companies want the public to show more interest before they invest in more recycled paper mills. The American Paper Institute, the EPA, and the Institute of Scrap Recycling all point to lack of consumer demand as the main factor limiting the recycling of paper. We can help break this standoff by using as much recycled paper as is currently available, and showing manufacturers there is a market (see **PAPER**).

Plastics

While the plastics industry spends millions of advertising and public relations dollars trying to convince the public that plastic is recyclable, according to the EPA in 1990 only about 1 percent of recyclable plastic was actually being recycled. The manufacturing plants simply aren't in place yet to process more.

Even when local recycling programs accept plastics, it doesn't mean they are actually recycled. Often recyclers don't know what happens to the plastic once it leaves their center. In 1992, over 200 million pounds of plastic waste from recycling programs was exported from the United States to almost thirty foreign countries in Asia, South America, and Africa. A Greenpeace investigation of recycling facilities in Asia revealed that much of the plastic waste was not recycled at all, but simply dumped in landfills or in random locations. Plastic that is recycled is processed under conditions that endanger workers and the surrounding environment.

There are many problems with plastic recycling, and difficulties for manufacturers of recycled plastic products in achieving a consistent supply. This makes it hard to establish a regular market. The recycled plastics industry is in its

PLASTIC RESIN IDENTIFICATION SYMBOLS

Resin Identification Symbols from the Society of the Plastics Industry are a three-sided triangular arrow with a number in the center and letters underneath that represent the type of resin used:

PET or PETE: polyethylene terephthalate

The most commonly recycled plastic. Lightweight and transparent, it is used to make two-liter soda bottles and plastic liquor bottles. It is recycled into many products including bottles for cleaning products and other nonfood items, egg cartons, fibers (carpet yarns, paintbrush bristles, twine, rope, scouring pads, and fiberfill for pillows, insulated vests, ski jackets, and sleeping bags), industrial strapping, engineering plastics, and automobile distributor caps.

HPDE: high density polyethylene

Most commonly used to make milk and juice jugs. Recycled, it is made into lumber substitutes (to build shipping pallets, livestock pens, outdoor furniture, and signs), base cups for soft drink bottles, flowerpots, pipe, toys, pails and drums, traffic barrier cones, kitchen drainboards, bottle carriers, and trash cans.

V: vinyl/polyvinyl chloride (PVC)

Used to make flooring, shower curtains, house siding, garden hoses, and many other products. Not currently recycled.

LDPE: low density polyethylene

Used to make cellophane wrap, disposable diaper liners, and some squeeze bottles. Not currently recycled.

PP: polypropylene

Used to make packaging, pipes, and tubes. Not currently recycled.

PS: polystyrene (commonly called Styrofoam)

Used to make disposable coffee cups, take-out food packaging, egg cartons, and packaging "peanuts" pellets. Recycled in some areas and made into the same types of polystyrene products, insulation, plastic "wood," and hard plastic products such as rulers and pens.

OTHER: all other plastic resins and multilayer materials

Not currently recycled.

infancy and may not be viable in the long run—every time a plastic is recycled it goes to a lower grade and still doesn't break down in the environment even after milk bottles have been turned into park benches and flowerpots. While plastics recycling can help reduce plastic in landfills, it is not a sustainable activity as it requires a consistent input of nonrenewable crude oil and creates a consistent output of ultimately nonrecyclable and nonbiodegradable products.

Many plastic products now carry a symbol that is a twist on the popular chasing arrows recycling symbol, although it is not intended to constitute a claim of either recycled content or recyclability. Rather the symbols are simply a shorthand method of identifying the type of plastic the bottle or container is made from to aid recyclers in sorting these items in the event that they actually are recycled. The resin code is found on or near the bottom of the bottle or container (see sidebar).

While this is a voluntary program, more than half the states have adopted plastic legislation based on this system: Alaska, Arizona, Arkansas, California, Colorado, Connecticut, Delaware, Florida, Georgia, Hawaii, Illinois, Indiana, Iowa, Kentucky, Louisiana, Maine, Maryland, Massachusetts, Michigan, Minnesota, Mississippi, Missouri, Nevada, New Jersey, North Carolina, North Dakota, Ohio, Oklahoma, Oregon, Rhode Island, South Carolina, South Dakota, Tennessee, Texas, Virginia, Washington, and Wisconsin.

California requires that plastic trash bags of 1.0 millimeter or greater thickness contain 10 percent recycled post-consumer plastic. By January 1, 1993, bags of .075 millimeter or greater thickness must be 30 percent recycled post-consumer plastic. In addition, starting in 1995, all rigid plastic containers between eight ounces and five gallons in size sold in the state must meet at least one of four criteria:

- Be made from at least 25 percent post-consumer material
- Have a recycling rate of at least 25 percent (56 percent for PET)

- Be reusable or refillable
- Be a source-reduced container.

In Part Two, see **Clothing, Furniture, Pillows, Rugs and Carpets, Trash Bags.**

Steel

Steel is used for a wide variety of industrial, construction, and consumer products. The largest use by far for consumer products is to make auto bodies and "tin cans" (actually made of steel with a very thin protective coating of tin) although it is also used to make appliances, cutlery, razor blades, and other items.

Though not usually labeled as such, virtually all steel products made in the United States have some recycled content, made by one of two processes. The electric-arc method, used for most industrial steel, can, and usually does, use 100 percent recycled scrap. The oxygen-furnace method, used for most steel cans, containers, and auto parts uses about 25 percent scrap, along with virgin materials.

According to the Steel Can Recycling Institute, steel is America's most recycled material—more than 66 percent of all the steel in the United States is already recycled into new products, over 100 billion pounds each year. More steel is recycled than paper, glass, aluminum, plastics, and other metals combined and doubled. Scrap steel from auto bodies, household appliances such as refrigerators and ovens, ships, railroad cars, and industrial equipment has been recycled for years, so the major recycling effort is now concentrated on steel cans. More than 92 percent of all metal food containers sold in the United States are made from steel, yet cans account for only one quarter of the steel recycled. As of 1991 the recycling rate for steel cans was 34 percent.

Most steel can manufacturers in America belong to the Steel Can Recycling Institute. The average steel can contains about 25 percent total recycled content. Scientific Certification Systems has verified the recycled content of the cans made by SCRI members to contain a minimum 10 percent post-consumer content (in reality it is 11 to 18 percent). This certification does not appear on the can because it is an almost industry-wide verification. A product certification would only occur for an individual manufacturer.

When looking for steel products that might have recycled content, it is important to know the difference between the types of steel used to make consumer goods. Steel, by definition, is made primarily from iron, with a small amount of carbon (from coal) added as an alloy. Depending on the performance requirements needed, other materials are also added.

The type of steel used to make auto bodies, cans, cutlery, and woks is plain carbon steel. Made simply of iron with 0.1 to 1.2 percent carbon and even less manganese, this can be recycled and often contains about 25 percent recycled content whether it is so labeled or not.

Many products, however, are made from stainless steel which has a special ability to resist corrosion. The most extensively used type for household items is made from 71.95 percent iron, 18 percent chromium, 8 percent nickel, and 0.05 percent carbon. Stainless steel products can *not* be recycled and do *not* contain recycled material.

Tires

Only about 5 percent of the estimated 240 million car and truck tires discarded each year are currently recycled. The huge stockpiles, often in private junkyards, are highly susceptible to fires that cause tremendous air pollution.

The rubber in the tire (about 40 percent) can be recycled to make building materials, irrigation hose, and paving material, and recycled rubber can be added to virgin rubber and plastics to produce a variety of products such as laundry baskets, mud flaps, athletic and door mats, truck-bed liners, grocery-cart wheels, and retread tires. The remainder of the tire—steel belts, bead wire, and sidewall—can also be recycled.

RECYCLABLE

A recyclable product is one that can be recycled. But when considering the recyclability of a product there are two separate and distinct issues: Can a product or material technically be recycled?, and in real life can a product or material practically be recycled?

The difference between these two definitions of recyclable is the subject of hot controversy in the product/packaging labeling world. Some say that any product or material that can technically be recycled should be labeled "recyclable." Others say that the attribute "recyclable" should only be used on a label if, in fact, it can really be recycled in the community where the product or package is sold. While recyclables such as glass or newspapers can be recycled almost everywhere, other recyclables like plastics or motor oil can only be recycled in the limited number of areas where programs have been set up.

Whether or not it is economically feasible to recycle a product is yet another question. Often virgin raw materials are cheaper than recycled materials, thanks to government subsidies and policies. Transportation issues can also come into play—is it better for the environment to use a local raw material or recycled material transported from a great distance? The shipping costs of moving collected recyclables to a distant processing plant can eat up all the profit. And recently there have been gluts of recyclable material—especially newspapers—that become worthless when there are not enough buyers of recycled products. All these problems can be worked out. It's important now to support the market for recyclable material by buying recycled products, and not simply give up on recycling as impractical.

Federal Guidelines

There are no federal regulations for using the word *recyclable*; however, the Federal Trade Commission has given guidelines for using the word *recyclable*.

The FTC says "A product or package should not be marketed as recyclable unless it can be collected, separated or otherwise recovered from the solid waste stream for use in the form of raw materials in the manufacture or assembly of a new package or product. Unqualified claims of recyclability for a product or package may be made if the entire product or package, excluding minor incidental components, is recyclable. For products or packages that are made of both recyclable and non-recyclable components, the recyclable claim should be adequately qualified to avoid consumer deception about which portions or components of the product or package are recyclable."

Later the guidelines specify that a recyclable product must be able to be recycled into a new, similar product (glass to glass, paper to paper, plastic to plastic, etc.). If the product can be used as fuel in a municipal incinerator to create power, this would not be considered recycling.

Here's one of their examples of an acceptable claim: "A product is marketed as having a 'recyclable' container. The product is distributed and advertised only in Missouri. Collection sites for recycling the container are available to a substantial majority of Missouri residents, but are not yet available nationally. Because programs are generally available where the product is marketed, the unqualified claim does not deceive consumers about the limited availability of recycling programs."

If collection sites for recycling the material in question are not available to a substantial majority of consumers or communities, but collection sites are established in a significant percentage of communities or available to a significant percentage of the population, the word *recyclable* could be used if there are additional qualifications. To avoid deception the claim should be qualified to indicate the limited availability of programs by stating, for example, "Check to see if recycling facilities exist in your area" or state something like "recyclable in the few communities with facilities for colored HDPE bottles." Other examples of adequate qualification of the claim include providing the number of communities or the population to which programs are available, or the approximate percentage of communities or the population to whom programs are available.

On the other hand, a soda bottle labeled "Recyclable where facilities exist" might be considered deceptive if recycling programs for material of this type and size are available in a significant percentage of communities or to a significant percentage of the population, but are not available to a substantial majority of consumers. To avoid deception, the claim should be further qualified as already recommended.

State Regulations

States have widely varying regulations for using the word *recyclable.*

In California in order to be labeled "recyclable" a product must be able to be conveniently recycled in every county in the state with a population over 300,000.

In Indiana a recyclable material or product is one which "can be redeemed or returned at an identifiable location for the purpose of returning the material to the economic mainstream in the form of raw material for new, reused or reconstituted materials which meet quality standards necessary to be used in the marketplace."

Northeast Recycling Council and the state of New York agree that in order to use the term *recyclable,* one of the following criteria must be met:

- 75 percent of the communities or 75 percent of the population in the state have municipal recycling programs for this material category
- The material category has achieved a greater than 50 percent recycling rate statewide
- The manufacturer, retailer, or distributor must demonstrate that it has achieved a statewide recycling rate of more than 50 percent by weight, for that product or package.

In New Hampshire recyclable products include but are not limited to post-consumer and/or pre-consumer glass food and beverage containers, plastic milk containers, plastic soft drink containers, newspaper, tin-coated cans and steel cans, aluminum, corrugated cardboard, and mixed office paper.

Rhode Island, on the other hand, *requires* recycling for aluminum, automobiles, coated unbleached kraft beverage carriers, corrugated cardboard, glass food and beverage containers, laser printer toner cartridges, newspaper, office paper, PET soft drink containers, steel and tinned-steel food and beverage containers, used lubricating oil, vehicle batteries, white goods, wood waste, telephone directories, leaves, and yard wastes.

Recyclable Products

What all this boils down to for consumers who want to buy a product or package they can recycle is that we can't rely on the label to tell us what we can and can't recycle. Many manufacturers are responding to these regulations by removing the word *recyclable* from their products and packages altogether. Other manufacturers call their products and packages recyclable regardless of the availability of a local recycling center.

The primary markets for recycled materials right now are aluminum and other scrap metals, glass, paper, and plastic. These are easily recognizable even without the word *recyclable* on the label.

Because recyclability of any product is so specific to your local community, it's important to know which products technically can be recycled, and which can practically be recycled where you live. I suggest that you take an hour or two to find out how and where your recyclables can be recycled.

Most communities in the United States now have some sort of residential recycling program in place, tailored to best suit local variables such as demographics, population density, and local markets for recyclables.

Curbside collection, where recyclables are put out in separate bins or bags next to the garbage can on garbage collection day, is the most convenient method and greatly encourages participation. Why not recycle when it's as easy as putting out the trash?

More common, though, are drop-off centers—large containers, "igloos," trailers, or waste bins marked with the kind of materials they accept. Even if there is no curbside recycling, more communities, even small ones, have some kind of recycling collection in the parking lots of supermarkets, natural food stores, or schools. Call your local garbage collection agency to find out where your local collection sites are.

If you live in a more populated area, you may have a local buy-back center, where cash is paid for recyclables that are brought in. A miniature version of the same idea can be found in some supermarkets, where they have a "reverse vending machine"—instead of putting money in and getting a can of soda, you put an empty can in and get back a nickel.

Many communities now have a recycling center at their local landfill. Ours collects various kinds of paper and cardboard, glass, cans, all kinds of scrap metal, lumber, paint, and miscellaneous items (dishes, books, auto parts, etc.) that can be reused.

Separating trash and recyclables is easy once you have a system. The Glass Packaging Institute estimates separation takes only fifteen minutes a week. It doesn't even take me that long—I have separate baskets for each type of recyclable item so instead of throwing something I can recycle into the trash, it goes straight to the appropriate basket.

RECYCLABLES REMINDER

ALUMINUM AND STEEL CANS

Type *Can be recycled at (location):*

❏ Aluminum cans ______________________________

- ❏ Foil OK
- ❏ Pie plates, frozen food trays, etc. OK

❏ Steel cans ______________________________

Preparation:

- Rinse briefly to remove food or beverage residues;
- Check with your local recycling center to see if they prefer flattened or unflattened cans
- Remove paper or plastic labels

GLASS

Type *Can be recycled at (location):*

❏ Clear glass ______________________________

❏ Green glass ______________________________

❏ Amber glass ______________________________

Preparation:

- Include glass food and beverage containers only
- Do not recycle mirrors, drinking glasses, glass dishware, windows, glass cookware, light bulbs, or broken glass
- Rinse briefly to remove food or beverage residues
- Separate glass by color
- Labels and neck rings need not be removed, but do not include lids, corks, or other materials that are not directly attached to the glass.

PAPER

Type *Can be recycled at (location):*

❏ Newspaper ______________________________

❏ Corrugated cardboard ______________________________

- ❏ Brown paper bags OK

- ❑ Office paper ____________________
    - ❑ Laser-printed paper OK
- ❑ Mixed papers ____________________
    - ❑ Acceptable papers are: ____________________
- ❑ Glossy paper ____________________
    - ❑ Glued bindings OK

Preparation:

NEWSPRINT

- Include newspapers, comics, inserts only
- Tie in small bundles with twine or string or place in brown paper bags—no plastic

CORRUGATED CARDBOARD

- Flatten boxes
- Tie together with twine, or fill an open box with flattened boxes
- Remove any nonpaper material like plastic tape
- Check locally to see if you can also include brown paper bags

OFFICE PAPER

- Remove plastic windows and self-sticking adhesive labels from envelopes
- Include white paper only (check locally to see if you can include colored paper)
- Do not include fax paper, blueprints, or self-stick notes
- Do not include nonpaper items (paper clips and staples are OK)
- Check to see if your local center will accept paper that has been printed with a laser printer

MIXED PAPER

- Check with your local recycling center for types of paper accepted and preparation instructions

GLOSSY PAPER

- Include magazines and mail-order catalogs (some magazines have only glossy covers, which can be removed)

- Keep separate from all other grades of paper
- Check with your local recycling center to see if magazines with glued bindings are accepted

PLASTIC

	Type	*Can be recycled at (location):*
❑	1 PET or PETE	______________________
❑	2 HDPE	______________________
❑	6 PS	______________________

NOTE: Other plastics are not yet recyclable, even though they may be labeled as to their resin type.

Preparation:

- Empty containers and rinse briefly to remove beverage residues.
- Remove metal lids (paper labels are OK)
- Flatten bottles

OTHER

	Type	*Can be recycled at (location):*
❑	Batteries (home)	______________________
❑	Batteries (car)	______________________
❑	Motor oil	______________________
❑	Paint	______________________

C H A P T E R 7

Sustainably Harvested

In order for a product to be truly sustainable, the resources from which it is made must be taken from the earth in a way that will allow the resource to renew itself and continue to be available to us for future use. "Sustainably harvested" means just what it suggests: A renewable resource has been harvested in a way that allows its inherent regeneration and continued supply.

While any renewable resource—plants, animals, surface water, soil, and sunshine—can be sustainably harvested, commercially, sustainably harvested is currently only being used with reference to forest products—from both domestic forests and the rain forests of the world—and this chapter will focus on these.

A standing forest performs functions vital to the great ecosystem of the earth. Forests create soil, moderate climate, control floods, and store water against drought. They cushion the erosive effect of rainfall, hold soil on slopes, and keep rivers and seacoasts free from silt. They harbor and support most of the earth's species of life. Forests take in and hold a great stock of carbon, which helps balance the stock of carbon dioxide in the atmosphere and thus combats the greenhouse effect.

Over the last ten thousand years, the earth's forests and woodlands have shrunk by one third as the result of humans clearing trees to make way for crops, pasture, and cities. For many centuries, forests were cleared to provide open land on which expanding populations could build, and the fallen timber was used as a locally available material for building. Then the surrounding forests were used for heating fuel, and in this way, we have lost many millions of acres of primal forest.

Before humans invented agriculture, there were 15.3 billion acres of forest on earth. Now there are 10.3 billion, only 3.7 billion of which are undisturbed primary forest. Half of that forest loss has occurred between 1950 and 1990. As the demand increases for timber, paper products, and fuelwood, forests in nearly every part of the world are disappearing. Virtually all Europe's original forests are gone, replaced by managed tree stands. In the contiguous United States, less than 5 percent of primary forest is left.

The United States harvests about 17 billion cubic feet of timber each year, and imports roughly another 2 billion cubic feet, primarily from Canada. Twelve billion cubic feet (63 percent!) is used for paper and pulp (another reason to recycle paper), leaving 7 billion cubic feet to be used for lumber.

About 2 percent of the wood we use in the United States comes from rain forests, according to the Rainforest Action Network. This sounds like a small amount, but our annual trade in tropical timber products is around $2 billion. Most tropical timber imported here is in the form of plywood, paneling, and veneer produced in Indonesia, Malaysia, Taiwan, and other Southeast Asian countries, although we also import some hardwood logs and manufactured items such as furniture, tableware, and musical instruments. The rest of the rain-forest wood goes to other countries that have already demolished their own forests, particularly Japan.

Tropical rain forests cover less than 2 percent of the globe, yet their lush ecosystems support nearly 50 percent of all species of life. Twenty-five percent of our medicines contain active ingredients derived from rain-forest plants and many common foods and spices originated in the rain forest including allspice, avocados, bananas, black pepper, cashews, chiles, oranges, rice, and vanilla. Brazil nuts can be grown only in healthy rain forests. Yet only a small percentage of rain-forest plants have been evaluated for potential uses—about three thousand fruits alone could be added to our diets.

The latest data from the World Resources Institute show that destruction of tropical forests is occurring nearly 50 percent faster than previous estimates, with enough trees being lost each year to cover the state of Washington. New satellite measurements show between 40 million and 50 million acres being lost each year. Deforestation results in water pollution, erosion, flooding, drought, loss of habitat and species, and displacement of native peoples.

Common products which destroy the rain forest include tropical wood products (there are more than seventy different woods that come from the rain forest—teak, mahogany, rosewood, and ebony are the most popular), imported beef (usually sold as fast food and in less expensive products such as luncheon meat, pet food, baby food, sausage, and frozen dinners), and virgin paper products (many American mills import tropical wood pulp). Disposable chopsticks are made from rain-forest woods—bring your own when going out for Chinese or Japanese food, or choose a restaurant that has reusable chopsticks.

SUSTAINABLE FORESTRY

Wood is a renewable resource, but sustainable only if forests are maintained and no more trees are cut than the forest can regenerate. Ninety-eight percent of the wood used in the United States is from our own domestic forests, which we are clear-cutting at a rate faster than they can naturally replenish themselves. Softwood forests on the West Coast of the United States during the 1980s exceeded sustainable yield by 25 percent on industry-owned land and by 61 percent in government-owned national forests. Logging in Canada's province of British Columbia in 1989 was 30 percent higher than sustainable yield.

The major reason that our forests are being overharvested is to keep up with consumer demand. Instead of saying "consumers demand X amount and we need to produce it regardless of environmental impact," the sustainable guideline would be "our forests can sustainably produce X number of board feet per year and that is what is available to consumers." We could find find ways to live with that. Harvest rates could be greatly reduced by eliminating waste and increasing recycling of paper and wood products. Efficiency of materials use could be increased through clever designs that use standard paper and cut-wood sizes.

When wood is needed, the first choice could be reclaimed (see Chapter 6) wood, such as lumber from demolished buildings. Instead of dynamiting buildings, they should be taken apart piece by piece for salvage. Scrap wood can also be used to make many products, as can fallen wood and twigs.

Most wood sold in lumberyards and used to make products is produced and harvested by the clear-cut–replant method. On the surface, the idea of clear-cutting and replanting makes some sense. Foresters are quick to point out that clear-cutting and replanting is sustainable and roadside signs next to young-growth trees note when the last cut was made and when the next harvest will be. In fact, a common argument in favor of the clear-cut–replant method is that we have more trees now than in the past because of tree planting. The numbers may be true, but the age, strength of the wood, and quality of the forest are not the same. Clear-cutting and replanting on a large scale do not maintain the integrity of a natural forest, they simply clear the land for tree plantations and tree plantations are not forests.

In a natural forest the seedlings of mixed native species replace old and dying trees. There are layers of plants—such as ferns, shrubs, saplings, and wildflowers—that grow under the forest canopy as well as birds, animals, and insects. Tree plantations are mainly composed of one or two fast-growing tree species that are often not native. Frequently any underbrush is cleared so as not to interfere with the tree growth or logging equipment.

When forests are clear-cut, the physical, chemical, and biological properties of the soil are changed. Instead of being nourished by the constant decay of forest flora, the soil is now exposed to erosion from wind and water. Wildlife are scattered to other nearby forests, where they can upset the delicate balance of populations. Vegetation that thrived in the forest often cannot grow without the protective forest cover. Rainfall runs off the soil, flooding nearby streams instead of sinking slowly and deeply into underground water tables.

Replanting programs are often unsuccessful. Trees are accustomed to growing in an atmosphere of mixed species, and do not thrive in single species-plantations. Soil becomes depleted and cannot nourish the trees. Artificial fertilizers, pesticides, and herbicides become a necessity. In certain areas defoliants such as Agent Orange have been used to destroy viable trees in lots simply because they are "the wrong species." Tree diseases and pests run rampant in single-species plantations.

Sustainable forestry—also sometimes called "restoration forestry," or "eco-forestry"—harvests some trees and leaves the forest. The general philosophy is to use the natural functions of the forest ecosystem as a model; however, as of yet there is no generally accepted name or definition that is commonly agreed upon.

Different interpretations of sustainable forestry incorporate one or more of these principles:

- Mixed species and ages—a variety of complementary species with trees of different ages growing together
- Selective cutting—trees are taken one by one, not clear-cut
- Sustainable yield—no more trees are taken than the forest can naturally replenish
- Soil conservation—care is taken to preserve and enrich the soil
- Ecosystem preservation—care is taken to maintain the natural flora and fauna
- Appropriate technology—nonpolluting tools are used such as two-man saws and horse carts, or power tools using nonpolluting fuels.

A beautiful description of this kind of forestry is given in the book *Wildwood: A Forest for the Future* by Ruth Loomis with Merv Wilkinson:

> My 136 acre tree farm, "Wildwood," on Vancouver Island, is a sustained yield, selectively logged tract of timber that has been producing forest products since 1945, and will continue to do so indefinitely. Here, there is still a forest and it is growing faster than I log it. I make a good living without destroying the forest. I work with

> nature. I harvest trees periodically for specialized products and on a regular basis for lumber, enjoying the forest, and its tranquillity where all the living organisms are present and healthy . . .
>
> I must emphasize *sustainable.* Selective is not good enough. I've met loggers who claim to do selective logging. This usually means that the best trees will be selected for cutting and the rest left with little consideration for regrowth or future production . . .
>
> Sustainable means "to sustain" or, to keep the trees in production, sustaining the level of growth of the stand and never over-cutting that growth. These 136 acres grow between 500–700 board feet per acre per year which gives a potential harvest of 68,000 board feet of timber each year. . .
>
> Harvesting of the trees can be done in cycles or continuously as long as the proper percentage is taken. I never cut over the annual growth rate, as that breaks into the "bank account" of the tree farm . . . I have now learned to leave five percent of my "interest," or annual growth, to decay and rot on the forest floor, a reinvestment in the soil, a reinvestment for the future . . .
>
> There are fundamentals I follow on Wildwood Tree Farm. I know the annual growth of the trees. I understand species-value according to the land and terrain. I thin for light and growth to encourage the proper canopy. I let reseeding occur naturally. I must take care of the soil, making sure enough wood debris is returned; and I consider carefully the factors of road building and erosion . . . The common denominator for good sustainable logging is to work *with* nature, using good common sense . . .
>
> A forest depends on the quality of its soil. The soil consists of many things: the organic matter and plant nutrients; the sub-soil underneath which provides the trees with anchorage and mineral intake; the bacteria and micro-organisms; the fungi and the bugs and critters that aerate the soil. It is a balanced entity . . . if you destroy that balance, you're going to be in trouble . . .

As part of my research into products, I once received a letter from a hardwood mill that said "Unlike west coast sawmills, mills in the northeast do not engage in the clearcutting of forest stands, but go into mixed species wood lots and remove only the mature trees, leaving all of the immature growth for future harvesting. There is virtually no seeding of replacement trees; the woodlots are naturally replenishing as all of the seedlings and unmature trees are left in the woods. While this is probably much less economically productive, it is the way mills and loggers have done it in the New England states for a long time. Most woodlots have as many as a dozen different hardwood species as well as many softwood species growing on them and the loggers are accustomed to selling more than one species of log even on each trailerload they deliver to the mill. Many of the private woodlot owners employ a part-time forester to mark the trees to be harvested." I've had this confirmed by others who know wood practices in the area. Generally hardwoods are selectively taken for furniture and lumber, but clear-cuts still occur for the softwood used to make pulp for paper manufacturing.

Sustainable harvest from the rain forest means different things to different people. The Rainforest Action Network warns us not to confuse "sustained-yield forestry," which is the practice of removing only as many trees as the forest can regenerate, with "sustainable forestry," which aims to maintain the integrity of the entire forest:

> The notion of "sustainable forestry"—forestry that meets the needs of the present without threatening the needs of the future—has been much discussed in recent years. The tropical timber industry, in particular, is quick to claim that it's preserving tropical rainforest by practicing "sustained-yield" forestry. But sustained-yield forestry is by no means the same thing as forestry that manages a forest in such a way that its ecological systems retain their integrity.
>
> In practice, sustained-yield forestry means the removal of the maximum possible number of trees in a given area without affecting the forest's ability to generate a given volume of wood. As such, it fails to consider the impact of logging on such critical ecological features as the forest's ability to maintain biodiversity and its role in hydrological cycles and climate regulation. At the same time, sustained-yield forestry also fails to consider the impact on the indigenous people who live in a forest, and their dependence on such forest products as nuts, fruits, bush meats, resins, and water sources.
>
> In reality, for all intents and purposes, there are practically no tropical logging operations anywhere in the world that can be considered sustainable . . . Truly sustainable timber operations can be found in less than one-eighth of one percent of rainforest lands . . . However, in the last few years, as the demand for truly sustainable tropical timber has increased, a number of promising experiments have gotten underway.

Their recommendation is to avoid the use of tropical hardwoods entirely, and instead use American-grown temperate woods and wood products.

Certification Organizations

A handful of certification programs are in development and some have certified small tracts of forest. Creating standards for wood certification is more difficult than creating a standard for, say, recycled content, where the product is made in a factory and you can verify that you collected so many tons of recycled material and your product has a certain percentage of recycled content. Wood is a product of the natural forest, not a factory, and so the requirements are very site-specific. Sustainable forestry is as much an art as a science and a sustained yield of wood is the successful result of knowing one's land and acting appropriately on it. Therefore certifications must be given based on the fulfillment of a basic philosophy that leads to a desired balance, rather than a strict set of do's and don'ts.

In March 1992 foresters, certifying organizations, and others involved in forest management worldwide formed the Forest Stewardship Council (FSC). The aim of this organization is to encourage good stewardship of all types of forests internationally—temperate, tropical, and boreal. It promotes forest management that is environmentally appropriate, socially beneficial, and economically viable to ensure that timber and nontimber products are harvested to maintain the forest's biodiversity, productivity, and ecological processes as well as provide an incentive for the local people and society at large to participate in "sustainable" management.

The basic purpose of the FSC is to provide guidance to consumers through the accreditation and monitoring of certification programs. Its "Principles and Criteria of Forest Management" (see sidebar) will be used by regional certifying organizations that will develop more detailed standards appropriate for local conditions. By this process it hopes to decrease consumer confusion by making the basic philosophy behind the standards consistent from group to group.

All certification organizations mentioned in this chapter are part of the Forest Stewardship Council.

Scientific Certification Systems

Scientific Certification Systems has a Forest Conservation Certification Program that developed out of its original certification program for sustainably harvested wood. It has moved away from calling its wood "sustainably harvested" because it believes the term is misleading; consumers might think the forest is guaranteed to be sustainable into the future and that guarantee is not possible. Because the natural workings of the forest are complex and we have much to learn, it can certify only the methods in practice today and just speculate on the future.

Instead of giving a single certification that meets a standard, the SCS certification is more like its Environmental Report Card with many factors considered (see Chapter 3). Certification is given based on the professional subjective opinion of a multidisciplinary team that evaluates three areas: sustainability of timber resources (defined as "continuous yield of merchantable timber over time"—a complete clear-cut wouldn't qualify), maintenance of the forest ecosystem, and socioeconomic benefits to the surrounding community (see sidebar). For each category, a set of evaluation criteria is established, with a score given for each category (100 being completely sustainable).

The first domestic wood to be certified by Scientific Certification Systems as being "harvested from a State-of-the-Art Well-Managed Forest" was that grown on the Menominee Indian Reservation in Wisconsin. Their reservation contains eleven of the sixteen major types of forest habitat in Wisconsin and more than twenty-five species of timber. All the original species of timber still flourish there, with the single exception of elm.

FOREST STEWARDSHIP COUNCIL "PRINCIPLES AND CRITERIA OF FOREST MANAGEMENT"

Principle 1. Management Plan: A written management plan must exist that clearly states management objectives for each forest and the means for achieving those objectives, and provides for responses to changing ecological, social, and economic circumstances.

Principle 2. Forest Security: The ownership of the forest must be clearly defined and documented, and management areas dedicated by the owners to permanent forest cover.

Principle 3. Social and Economic Benefits: Participating parties should receive an equitable share of the benefits arising from forest production activities.

Principle 4. Local Rights: The legal and/or customary rights of indigenous peoples and other long-settled forest-dependent communities affected by forestry activities must be protected, and forest management planning and implementation must provide for full and informed consent in relation to activities that affect them.

Principle 5. Environmental Impact: Forest management activities must have minimal adverse environmental impact in terms of wildlife, biodiversity, water resources, soils, and nontimber and timber resources.

Principle 6. Sustained Yield: Harvesting rates of forest products must be sustainable in the long-term future.

Principle 7. Maximizing the Forest's Economic Potential: Forest management should take into account the full range of forest products—timber and nontimber—and forest functions and services, and should maximize local value-added processing.

Principle 8. True Costs: The cost of forest products should reflect the full and true costs of forest management and production.

Principle 9. Appropriate Consumption: Forest production should encourage judicious and efficient use of forest products and timber species.

Principle 10. Forest Plantations: Plantations should not replace natural forest; they should explicitly augment, complement, and reduce pressures on existing natural forests.

Principle 11. Chain of Custody: The original sources and subsequent steps in the processing and supply chain must be documented to allow accurate product tracing.

About 75 percent of the forest is managed sustainably. Selective cutting is used to weed out trees that are damaged, slow-growing, or stunted by disease or competition. Only if all the inferior trees have been taken and the stand is still too dense will the more valuable timber trees be cut. A system of monetary fines controls loggers from doing excessive collateral damage to the forest, or taking too many trees. But the practices aren't perfect. Roughly 20 percent is managed for aspen and jack pine with a clear-cut program, and less than 5 percent is tree plantation, planted after the clear-cuts. "Prudent" applications of herbicides are also used.

Also certified is the wood from the Collins Almanor Forest, a 92,000-acre forest in northeastern California harvested by the Collins Pine Company. The family-owned business was acknowledged for demonstrating "consistent appreciation for and cooperation with the natural forest processes." The majority of lumber products made from this wood are high-quality wood molding, window components, and door parts; other products go to furniture makers, home-improvement centers, and home builders. At its mill in Chester, California, waste tree bark and sawdust are used to generate steam and electricity.

The one tropical wood source certified by SCS is the Plan Piloto Forestal program in Mexico. There the impetus for sustainable management comes from its communal effort and interest. Two communities have held their forest lands for fifty-five years, with their economy based on the sale of resins from their chicle trees (used in chewing gum) and sustained yield of mahogany, Spanish cedar, and other valuable woods. They are committed to sustainable forestry management as they know it is the only way to provide themselves with a solid, reliable livelihood.

SCS is working to certify other forests under their Forest Conservation Program. Look for its label on both lumber and products made from wood.

Pacific Certified Ecological Forest Products

A more grass-roots program is Pacific Certified Ecological Forest Products (PCEFP). A project of the Institute for Sustainable Forestry, this program grew out of woodworkers', consumers', and activists' growing interest in lumber produced without harm to the forest.

To devise its standards for ecologically produced wood, the Institute drew on the experience and knowledge of ecologists, foresters, fisheries biologists, watershed restoration workers, fallers, planters, and numerous other forest workers and experts.

The Institute considers for certification wood and other forest products harvested in the redwood and Douglas fir hardwood forest types of northern California and southern Oregon, but may expand to other areas as experience and need permit. More important, this program is an excellent model that can be implemented locally in other areas, where those who know their forests best can develop and execute their own programs.

The criteria for certification are founded on several philosophical and ecological premises: a belief that forests are whole living fabrics, greater than the sum of their component species, physical elements, and human practices and

SCIENTIFIC CERTIFICATION SYSTEMS FACTORS FOR EVALUATING A "WELL-MANAGED FOREST"

Timber Sustainability

This element focuses on the management of the timber resource and the extent to which the management unit's productive potential for yielding high-quality timber products is being realized by current silvicultural prescriptions and management regimes. To be qualified for evaluation under this program element, the forest ownership must have a formal, written timber management plan that is updated at regular intervals. Criteria considered in this element include

- rate of harvest vs. growth
- rotation lengths
- site productivity
- harvest efficiency.

Forest Ecosystem Maintenance

This element focuses on the extent to which the management of the timber resource alters and/or adversely impacts the natural forest ecosystem, thereby reducing or eliminating key ecosystem elements such as fish and wildlife habitat and other beneficial uses of water. Criteria considered in this element include

- serial distribution (the distribution of different ages of trees in the forest)
- watercourse and watershed management
- retentions (areas of ecological importance that are set aside and not harvested)
- growth and management of the timber
- logging and road construction
- wildlife management.

Socio/Economic Benefits

This element focuses on the various social and economic benefits derived by the surrounding community as a result of the forest management operation. It also examines the financial aspects of the operation itself. Criteria considered in this element include

- financial stability
- socio-economic relationship with the community
- employee relations and practices.

that the health of the forest takes priority over human needs for products to be extracted from the forest. Without intact forests, we can have no forest products.

Because forest conditions and appropriate practices are extremely site-specific, the Institute has issued forest practices guidelines instead of stringent standards. These are based on its "Ten Elements of Sustainability" (see sidebar).

Landowners wishing to have their wood certified as ecologically harvested must first prepare a Forest Management Plan with an Institute-certified Registered Professional Forester. This document outlines the plans for stewardship and harvest of the forest, and explains how the practices satisfy the Ten Elements of Sustainability. Certifying inspectors visit the site before, during, and after the harvest, and may revoke certification if the plan or its spirit are disregarded. Consumers are encouraged to use the certified wood only for finished products that will last at least as long as the age of the trees that were cut to supply the wood.

Rainforest Alliance Smart Wood Program

The Rainforest Alliance is a nonprofit conservation organization that takes a multifaceted approach to preserving the rain forest. One program purchases land to add to rain forest reserves, another identifies and promotes the use of wood and wood products whose harvest does not contribute to the destruction of forests. Since 1990 the Smart Wood certification program has certified wood and wood products which come only from sustainable or well-managed forestry operations in the tropics.

PACIFIC CERTIFIED ECOLOGICAL FOREST PRODUCTS "TEN ELEMENTS OF SUSTAINABILITY"

1. Forest practices will protect, maintain and/or restore the aesthetics, vitality, structure, and functioning of the natural processes, including fire, and of the forest ecosystem and its components at all landscape and time scales.
2. Forest practices will protect, maintain and/or restore surface and groundwater quality and quantity, including aquatic and riparian habitat.
3. Forest practices will protect, maintain, and/or restore natural processes of soil fertility, productivity, and stability.
4. Forest practices will protect, maintain and/or restore a natural balance and diversity of the native species of the area, including flora, fauna, fungi and microbes, for purposes of the long-term health of ecosystems.
5. Forest practices will encourage a natural regeneration of native species to protect valuable native gene pools.
6. Forest practices will not include the use of artificial chemical fertilizers or synthetic chemical pesticides.
7. Forest practitioners will address the need for local employment and community well-being and will respect workers' rights, including occupational safety, fair compensation, and the right of workers to collectively bargain and promote worker owned and operated businesses.
8. Sites of archaeological, cultural, and historical significance will be protected and will receive special consideration.
9. Forest practices executed under a certified Forest Management Plan will be of the appropriate size, scale, time frame, and technology for the parcel, and adopt the appropriate monitoring program, not only in order to avoid negative cumulative impacts but also to promote beneficial cumulative effects on the forest.
10. Ancient forests will be subject to a moratorium on commercial logging during which time the Institute will participate in research on the ramifications of management in these areas.

All sources of tropical timber are potentially eligible for evaluation and certification, including natural forests and plantations, large commercial concessions, and small community-based forestry operations. Before giving a certification, Smart Wood conducts field investigations of sources and companies using the Smart Wood guidelines. These are based on three broad principles: maintenance of environmental functions, including watershed stability and biological conservation, sustained-yield production, and positive impact on local communities. Professional foresters, ecologists, and social scientists have participated in developing the guidelines, which are constantly revised and improved. Smart Wood is also developing standards for specific regions and countries in cooperation with local people and governments.

Wood sources are certified on two levels. A sustainable source operates in strict adherence to the guidelines; a well-managed source had demonstrated "a strong operational commitment" to the guidelines. Companies that sell Smart Wood products can also be certified. As of May 1993 the program has certified five production sources and thirteen companies supplying wood or wood products from these sources.

Cultural Survival

Roughly 200 million indigenous peoples in two thousand known forest tribes live in tropical rain forests. Before the forests can be clear-cut, these people must first be displaced, resulting in disease, poverty, and dependency of formerly self-sufficient peoples living sustainably in the forest.

Cultural Survival has a program to market sustainably harvested rain-forest products in affluent countries, designed to provide a sustained income for rain-forest peoples and enable rain-forest residents to profit fairly from its efforts, instead of marketing its products through middlemen. The purpose is to show that the living forests can generate more income and employment than the same areas cleared and used for other purposes. A 1989 study by the New York Botanical Garden's Institute for Economic Botany showed that a rain forest actually is worth more intact than logged or burned for cash crops or cattle pasture: The fruits and rubber that can be harvested from a hectare of rain forest can be sold for $455 a year, every year, which is more valuable than selling it for logging or burning once for $1,000.

Cultural Survival purchases forest products for international sale from undisturbed forests, sustainable systems of agroforestry, and reforested areas on degraded lands. Profits from product sales support existing organizations of rain-forest peoples working to ensure the political and economic growth of their communities. During the first year, enough money was earned to finance the first Brazil-nut processing plant that is owned by the nut collectors themselves.

On the surface this sounds like a good program—the picture that comes to mind for me is that if I buy this product, I help save the rain forest so that indigenous peoples can continue to live there according to their ancient, sustainable ways, but this isn't, in fact, what happens. True, this type of sustainable harvest does save the rain forest habitat if done correctly, but the profits serve to bring the native peoples into industrial commerce, rather than allow them to continue their cultural traditions. Many native peoples around the world do not want to be involved with or dependent on the sale of their resources to industrialized nations, and suddenly having newfound money has divided the social fabric of some native communities rather than bringing them together.

And there is some question about whether the rain-forest ingredients used in products are actually sustainably harvested. According to an article in *E Magazine* (July/August 1993), the former director of Cultural Survival, Jason Clay, has stated that the only way to develop sustainable sources is to buy first from the open market, even though that means that they don't know how the products were harvested, or by whom, or under what conditions. Gradually they develop sustainable sources—today only four of their fifteen initial commodities are still purchased from the commercial market. What they do guarantee is that the money generated from all sales goes back to set up a sustainably harvested system.

Many rain-forest products are not actually harvested from the rain forest. Cashews, a common ingredient in many rain-forest cereals and snacks, come from trees grown in plantations where rain forests once stood. Some of the commodities are purchased from colonists, not indigenous peoples, with the idea that if there are no other means for the colonists to make a living, they will keep cutting down the rain forest to sell for pasture.

The rain forest has produced many products over the years for local and international markets (particularly rubber and kapok), but the recent demand for rain-forest ingredients is leading to some unsustainable practices. Cupuaçu, for example, a fruit that tastes something like a honeydew melon, grows wild in the rain forest, can be sustainably harvested, and is being used as an ingredient in bottled juice. Because of the new market, many farmers are now planting cupuaçu to cash in. If too much cupuaçu is harvested, the price could drop, leaving farmers without support from the international market or their traditional ways of living.

Sustainably Harvested Products

There are few products currently available made from sustainably harvested wood, although in the future I suspect that all companies will have to adopt some kind of sustainable practices. If this claim is used, be sure to get detailed information to confirm for yourself that the practices are sustainable (see **FURNITURE, GARDEN, PENCILS**). Buy products made from domestic wood rather than wood from the rain forest.

I'm a little wary of the idea of saving the rain forest by marketing its products through industrial commerce. I question the sustainability of this for two reasons.

First, using sustainably harvested rain-forest ingredients in consumer products may in fact help keep the rain forests standing if they are sustainably harvested, but rarely are the remainder of the ingredients in these products produced in a sustainable way. The argument is that people are buying ice cream or shampoo anyway, so these are logical things to add rain-forest products to.

However, strictly from a sustainability point of view, there is an enormous difference between the sustainability of a candy with rain-forest nuts and simply eating the plain nuts. The candy might be 40 percent sustainable because of the other ingredients, whereas the plain nuts could be 100 percent sustainable. And is the fact that we are told we can save the rain forest encouraging us to eat more candy and cookies?

More importantly, though, is that building a sustainable world can only be accomplished by building sustainable communities. Commodities such as fruits and nuts and wood need to be sustainably harvested in the regions of the world where they are being used, not shipped in from far away. For centuries indigenous peoples sustainably managed their resources through local markets. To be sustainable, we should be sustainably managing our resources for our own use and allowing them to manage theirs for their use—we should not take the resources of other countries or make them dependent on our purchases for their sustenance. If there were no market for rain-forest woods or cheap hamburger, the rain forests would stay standing.

But the unanswered question is: In the short term, does selling rain-forest ingredients in consumer products really help keep the rain forest intact? If so, it may be sustainable to support these products for this reason alone. After all, once the rain forest is clear-cut or burned, there is no opportunity left for the sustainable harvest of anything. But I haven't been able to find any documentation that any rain forest actually has been saved as a direct result of our consumer purchases. So I can only fall back on nature and say: Eat what nature has provided for you in the place where you live.

CHAPTER 8

Energy-Efficient
Energy-Saving
Water-Efficient
Water-Saving

One of the basic principles of sustainability is to minimize our use of resources, and when we do use resources to make the most of them by using them efficiently. Our reserves of nonrenewable resources will last longer when used efficiently, and when we use renewable resources efficiently, we make it more possible for their rate of supply to fulfill our rate of need. There is no waste in nature; each organism and ecosystem uses the resources it needs in an efficient way, and no more. We should do the same.

Efficient use of resources is an issue both in the design and manufacture of products, and in our use of the product at home. Since efficiency in design and manufacture is not generally revealed on a product label, this chapter will focus on products that allow us to make efficient use of two vital resources: energy and water.

ENERGY

I remember very clearly when energy became a concern for America. I had just graduated from high school and my parents bought a car for me so I could commute to college. It was my dream car: a 1967 Pontiac Firebird Formula 400 with racing stripes and hood scoops. After a year of sitting in line at the gas station on an even-numbered day to buy my ration of gas, I finally couldn't drive it anymore because I couldn't buy gas for it—my Firebird only ran on 100 octane gas and it was no longer sold.

The energy crisis of the mid 1970s created an awareness of energy conservation that has resulted in the energy-efficient products of today. If we took all the known and available energy-saving measures, we could decrease tremendously the amount of energy used and the resulting environmental problems. Ultimately, creating a sustainable world will require developing a renewable energy system—using energy efficiently now, thereby reducing the amount of renewable energy needed, is our first step.

Throughout history our growing population and developing technology have resulted in an increasing need for more and more energy. The first energy needs were satisfied with our own muscle power, then fire was used, and animals. Later wind was captured to power ships, flowing water turned waterwheels at the flour mill, and the sun heated homes and provided hot water. One study estimated that in 1850, 94 percent of the total energy used in the U.S. manufacturing industry was human or animal; by the mid-1970s it had decreased to only 1 percent.

Today more than 90 percent of the energy used in the United States and around the world comes from nonrenewable fossil fuels (coal, oil, and natural gas), with a small amount from nuclear power. As long as we continue widespread use of these forms of energy, very few products can qualify as being truly environmentally sustainable.

The burning of fossil fuels for energy is perhaps the best example of an activity that is not sustainable. It depletes a nonrenewable resource and turns it into pollution. When fossil fuels are burned in motor vehicles, factories, and power plants, they turn into carbon dioxide, nitrogen oxides, sulfur dioxide, and a number of other combustion products which do not, on any time scale of relevance to humanity, come back together to form fossil fuels again. In addition, these air pollutants cause major environmental problems.

Nitrogen oxides are one of the necessary components in the creation of smog. Smog is now found in many cities and rural areas—it hangs over the Grand Canyon almost all the time. The EPA estimates that smog reduces crop yields by as much as 30 percent in some areas and costs billions of dollars in lost production.

Sulfur dioxide and nitrogen gases undergo a chemical transformation in the atmosphere to become acid rain (or acid snow, fog, or dew). Acid rain is now a problem all over the United States, with acidity of precipitation regularly five to ten times greater than normal. Southern California has experienced acid fog with acidity one to two thousand times normal. Forests, the maple syrup industry, agriculture, fishing, soil quality, and historic buildings and monuments have all already been damaged by acid rain.

Carbon dioxide and other air pollutants from burning fossil fuels contribute to climate changes. Many scientists are warning that we are headed for an increase in average temperature, called the greenhouse effect, in which the gaseous pollution we are creating causes heat to be trapped in the atmosphere. Others say this effect is temporary and that our pollution is accelerating an ice age. Whichever way it may go, major shifts in weather patterns could cause changes in agricultural and forest zones, severe droughts, and rising sea levels that could drastically flood coastal areas.

Nuclear power plants have created a major environmental problem that involves the disposal of highly radioactive wastes. While the Nuclear Waste Policy Act of 1982 calls for permanent storage in deep underground repositories, there is no guarantee against future leakage. Nor is there a guarantee that there will be no nuclear accidents, such as that which devastated Chernobyl in 1986. As with unsustainable fossil fuels, nuclear power mines nonrenewable resources from underground and produces wastes that cannot safely be assimilated back into the ecosystem.

The overwhelming majority of products in the marketplace today use vast amounts of energy in the production and harvest of raw materials, the manufacture of the product, its transportation to the consumer, and its disposal. Add the energy used for heating, lighting, and running machinery and equipment at the manufacturing plant, warehouse, and retail stores; the energy used for packaging and advertising; and the energy we as consumers use in our cars to go to the stores to buy the product, and the total energy cost to keep our whole system of commerce going becomes astronomical.

It is difficult to consider energy-efficiency when we are choosing products because in order to evaluate a product's real energy use, we would need to know how much energy was used in taking raw materials, in manufacture, transportation, use, and disposal (this energy used to make the product is called "embodied energy"). Embodied energy should be a prime consideration in choosing products because of its enormous impact on the environment, but because we have so little information on the type or amount of energy used to make any product, it can be difficult to include this in our decision-making process. In the future, when life cycle information is available for products, we will be able to compare the energy demands of one product to another.

The one area where we do have control over energy consumption is at the stage of the product life where we use it. Using energy to power our homes and automobiles utilizes about 30 percent of the total energy in this country. This is where we can choose to save energy and be energy-efficient.

Energy efficiency is the most cost-effective way to meet our energy needs, conserve resources, and reduce pollution. Buildings can be heated and cooled using a fraction of the energy now consumed; efficient motors can dramatically cut the energy needed to run them; lights and appliances can be made that require much less energy to run; cars could be made right now that would get more than one hundred miles per gallon. In fact, by using already existing energy-efficiency measures and known technologies, we can cut our total energy

BUYING LOCALLY PRODUCED GOODS

Another way to save energy is to buy products that are made close to where you live—in your town, county, state, or region. This saves the energy that would have been used to transport the product from its manufacturer to your local store, if the product originated from farther away.

A tremendous amount of goods are shipped around this country and imported from around the world, frequently unnecessarily. At my local supermarket I can choose from a brand of butter made from the cows that graze on the hillsides in the valley where I live in northern California, or a brand of butter shipped in from Wisconsin. If the only change you made in your buying habits was to buy locally produced goods, you would be taking a giant step toward improving our environment.

Here's an example of how locally made products can save energy. In a conversation with Scientific Certification Systems about their Environmental Report Card certification program, they showed me an inventory of all the materials used and waste created to make and deliver to the consumer a particular brand of recycled-paper toilet paper. They said the manufacture of this particular brand was state-of-the-art and showed me a bar graph that indicated a limited environmental impact. However, when the nonrenewable fossil fuel resources used and air pollution created in shipping this toilet paper from New England to California was added in, the bar graph indicated an environmental impact so great, that their conclusion was that there needed to be a factory in California to make recycled-paper toilet paper, to save the environmental impact of shipping. The right toilet paper for Vermont is not necessarily the right toilet paper for California.

The name of the manufacturer and place of manufacture is generally noted on the product label. For very large corporations, however, the city on the label is often the location of the company headquarters and not the location of the factory where the product was made (some products made by large corporations are, in fact, made in regional factories across the country). I have found, however, that there are often national and local brands that are easy to identify.

It's important to keep in mind, too, that the individual ingredients in the product are shipped in from many different places and this is not indicated on the label. But more and more local producers are joining together to market their local products as such. Just north of me in Sonoma County, there is a group of agricultural products that use a "Sonoma Select" seal, indicating that these products are grown and packaged in Sonoma County.

The sustainable ideal is to have at least most of our products made locally from local materials and sold to local residents. In addition to saving energy, buying local products help support your regional economy and encourages small business. And you can generally find out more information about what the product is made from and how it is made from a local producer than you can from a multinational corporation.

use in half or better (some estimates go as high as 80 percent) without affecting production or service.

ENERGY-EFFICIENT, ENERGY-SAVING

According to federal law energy efficiency is "the ratio of the useful output of services from a consumer product to the energy use of such product." It is a simple measurement, not an interpretation or judgment. Energy use is the "quantity of energy directly consumed at point of use," and does not include the amount of energy utilized in the manufacture or transport of the product and its ingredients.

Energy-saving and energy-efficient are two similar, but slightly different, terms used to describe a reduction in energy use and are not regulated by law. In common language and on product labels they are used interchangeably, but there is a difference, I believe.

Energy-saving simply indicates that energy that might have been used wasn't. Energy-saving implies a loss of service—if you turn off the lights when you aren't in the room you're saving energy but you also don't have any light, or if you use a 75-watt light bulb instead of a 100-watt light bulb you save energy, but the light is dimmer (although it may be perfectly adequate for your need).

Energy-efficient refers to using less energy without losing the service—a 75-watt energy-efficient light bulb gives the same amount of light as a 100-watt incandescent bulb, and energy-efficient refrigerators keep your food just as cold.

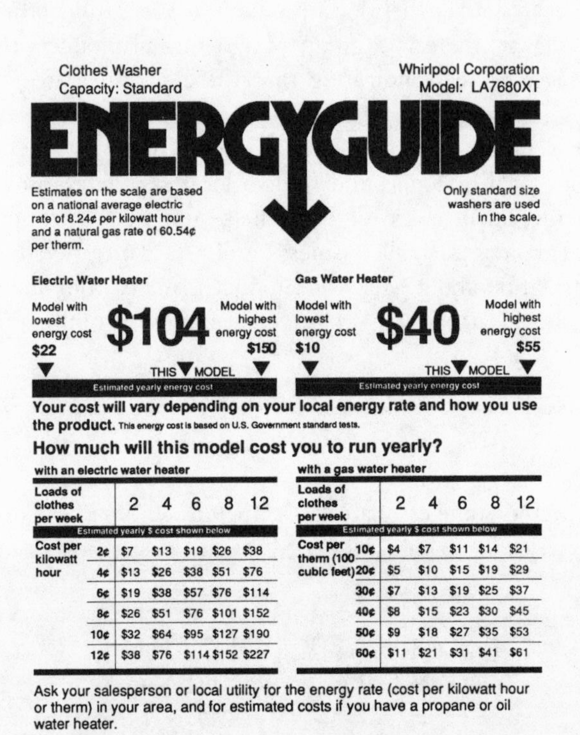

Clothes Washer
Capacity: Standard

Whirlpool Corporation
Model: LA7680XT

ENERGYGUIDE

Estimates on the scale are based on a national average electric rate of 8.24¢ per kilowatt hour and a natural gas rate of 60.54¢ per therm.

Only standard size washers are used in the scale.

Electric Water Heater

Model with lowest energy cost **$22** — **$104** — Model with highest energy cost **$150**

THIS MODEL

Estimated yearly energy cost

Gas Water Heater

Model with lowest energy cost **$10** — **$40** — Model with highest energy cost **$55**

THIS MODEL

Estimated yearly energy cost

Your cost will vary depending on your local energy rate and how you use the product. This energy cost is based on U.S. Government standard tests.

How much will this model cost you to run yearly?

with an electric water heater

Loads of clothes per week		2	4	6	8	12
Estimated yearly $ cost shown below						
Cost per kilowatt hour	2¢	$7	$13	$19	$26	$38
	4¢	$13	$26	$38	$51	$76
	6¢	$19	$38	$57	$76	$114
	8¢	$26	$51	$76	$101	$152
	10¢	$32	$64	$95	$127	$190
	12¢	$38	$76	$114	$152	$227

with a gas water heater

Loads of clothes per week		2	4	6	8	12
Estimated yearly $ cost shown below						
Cost per therm (100 cubic feet)	10¢	$4	$7	$11	$14	$21
	20¢	$5	$10	$15	$19	$29
	30¢	$7	$13	$19	$25	$37
	40¢	$8	$15	$23	$30	$45
	50¢	$9	$18	$27	$35	$53
	60¢	$11	$21	$31	$41	$61

Ask your salesperson or local utility for the energy rate (cost per kilowatt hour or therm) in your area, and for estimated costs if you have a propane or oil water heater.

Important Removal of this label before consumer purchase is a violation of federal law (42 U.S.C. 6302)

(Part No. 3356169)

The large number gives the yearly energy cost in dollars, based on estimated hours of use per year and a standard energy price. The standard energy price is an average for the entire country and changes over time as energy prices go up and down. Labels at different stores may have been printed at different times and thus may use different energy prices; therefore it is more accurate to find out your local price per kilowatt hour (mine is printed on my electric bill—$0.11439 this month) or therm of natural gas and use the yearly cost table below for comparison.

For some appliances, the large number is not dollar, but rather energy-efficiency ratings: EER for room air conditioners, SEER for central air conditioners, and HSPF and SEER for heat pumps.

NOTE: The presence of an EnergyGuide label on a product does not mean it is particularly energy-efficient—it simply gives a measurement of energy use that consumers can use for comparison.

There is no set percentage of energy reduction required for a product to be called energy-saving or energy-efficient. Each product is different and the individual technologies have to be considered. In general, though, these terms can legitimately be used if the product uses considerably less energy than other products of the same type conventionally used.

Another thing to keep in mind about energy efficiency is that to determine the real energy efficiency of a product we must also consider the energy used in the acquisition of that energy. The precombustion value of different forms of energy are very different—it takes a lot more energy to mine coal or drill for oil than it does to have water flow through a hydroelectric plant or for a solar collector to transform sunshine into power. In industrial life cycle assessments, this precombustion energy is included in the calculations because the combined total of the embodied energy and the precombustion energy represents the true energy demand of the system.

Electric power companies obtain energy from a variety of sources, including coal, nuclear power, hydropower, natural gas, petroleum, wind, solar energy, solid waste, and wood biomass. These fuels are interchangeable for the most part and utilities decide which to use at any given time based on which is most cost-effective at the time and availability. Nationwide in 1991 more than half our electricity came from burning coal and almost a quarter came from nuclear plants. Of course, this varies from place to place; where I live almost all our electricity comes from a nuclear power plant about seventy-five miles away. Just for your own awareness and information, you might want to find out from your local utility the source of the energy they are sending through your wires, and how much energy it takes to produce that energy. Multiply the amount of energy you use by the amount of energy it takes to produce that energy, and you'll have a more accurate idea of the amount of energy you really are using.

Federal Regulations

The federal government requires manufacturers to use standardized tests to determine the energy-efficiency of their products.

EnergyGuide labels—providing specific information about yearly energy costs and/or equipment efficiency—must be placed on all new refrigerators, freezers, water heaters, dishwashers, clothes washers, room and central air conditioners, and heat pumps. EnergyGuides are not required on kitchen ranges, microwave ovens, clothes dryers, heating equipment other than furnaces, "instant" water heaters, portable space heaters, lights, or many other products that use smaller amounts of energy, such as computers or calculators.

Energy-Efficient and Energy-Saving Products

Each of us in the United States consumes about seventy barrels of fuel oil or its equivalent each year, compared with a world average of only about eleven barrels. With only 5 percent of the world's population, we guzzle 25 percent of the world's energy supply. Energy consumption worldwide has continued to climb since the first oil well was drilled, and most of the energy flows through the industrialized world.

Although energy use varies depending on climate, here's a national-average breakdown of the way we use energy in our homes:

Space heating	32 to 60 percent
Space cooling	7 to 40 percent
Water heating (including clothes and dishwashers)	15 to 31 percent
Refrigeration/freezing	8 to 12 percent
Lighting	6 percent
Cooking	4 percent
Clothes drying	3 percent
Other appliances and electronics	6 percent

It is very, very important, however, for you to assess your own energy use and take action accordingly. My husband and I, for example, live in a relatively temperate area where the temperature ranges between 40 and 80 degrees most of the year. Even on the coldest winter nights it rarely gets down to freezing—last winter it got down to 32 degrees once and 28 degrees once, and a couple of winters ago we had an all-time record of 7 degrees during a highly unusual cold snap. In the summer we have a few days in the upper 90s, but mostly it's in the 70s and 80s.

So in our house we use very little energy for space heating and practically none for space cooling. We use scrap wood in our wood stove (my husband does carpentry and tree work and we have several years' worth of wood stacked up outside) and only on very occasional freezing mornings will I turn on a tiny space heater under my desk until the house heats up. We'll turn on a single small fan for a few hours in the evening maybe ten days out of the whole summer. Obviously our largest energy expenditure wasn't space heating and cooling, but it may be yours if you live in an area where there are greater extremes of temperature.

We had a big old energy-guzzling refrigerator that came with the house when we bought it, so we replaced it with a smaller, energy-efficient one that was the right size for a week's worth of perishables. And we took steps to save energy with water heating, and switched to energy-efficient light bulbs. We also eliminated the electric appliances we really didn't need. I use knives now to chop and slice instead of a food processor.

All energy-saving measures will pay for themselves over time and in some states qualify for tax rebates. Here in northern California, everytime I get my bill from Pacific Gas & Electric, there is another offer for an energy-savings rebate. Call your local utility and see if they offer in-home energy audits. Many utilities will come to your home or office for free (or at very low cost) and not only tell you how you can best save energy, but also let you know what kind of financial assistance is available, such as rebates and low-interest loans.

Here's an overview of what you can do to save energy at home. There are many books on the market that give tips on saving energy at home (see Recommended Reading for a few), so I'll just focus on issues that apply to choosing energy-efficient products.

Space Heating and Cooling

Reducing your home's energy use for heating and cooling can save more energy than anything else. Even though this isn't as simple as changing a light bulb, over the long run it's the most important.

The first step, however, isn't to run out and buy a more energy-efficient heater or air conditioner, but rather to weatherize and insulate your home. The greater the ability of your home to retain the heat or coolness, the less energy it will take to attain that ideal temperature. Air can escape from a building shell in hidden ways, so you might want to have an energy audit done by a trained professional.

INSULATION is the single most important product you can buy to reduce in-home energy use. It can save you up to 40 percent of your total energy use and often pays for itself in reduced costs within one to two years.

Hidden air leaks are among the largest sources of heat loss in older homes. **CAULK** and weatherstrip around doors, windows, cracks, and holes inside and outside. Putting gasket insulators behind all electrical outlets and switchplates can give you an additional energy savings of up to 10 percent or more.

About one third of a home's total heat loss usually occurs through windows and doors. As much energy leaks through our windows every year as flows through the Alaska pipeline, so increasing the insulative value of your windows can make a big difference.

To reduce heat loss, you can either fix or modify your existing windows, or replace them with new energy-efficient models (see **WINDOWS**). If your windows are generally in good repair, it will be more cost-effective and resource-efficient to improve their efficiency with weatherstripping and caulk. Exterior doors generally need only weatherstripping around the perimeter to ensure a tight seal when closed.

If you live in a cold climate you might already have storm windows, which can double the insulating ability of a single-pane window. The simplest storm window is plastic film taped to the inside of the window frame. This costs three to eight dollars per window and typically lasts one to three years. Removable or operable storm windows with glass or rigid acrylic panes make more sense environmentally from a resource use point of view and are more economical in the long run. The wooden or aluminum frame variety tend to be more energy-efficient.

Installing thick curtains or drapes on the interior of windows can also hold heat in, or even just hanging up blankets. Where I live we only have about a week or two of really cold weather, so around Christmas we just bring out the sleeping bags and hang them up on the bedroom windows—they keep our bedroom cozy and warm. Insulated shades or "window blankets" make your windows even more energy-efficient than a single layer of fabric (see **WINDOW COVERINGS**).

The most sustainable heat for your home is warmth from the sun. For centuries homes have been designed to take advantage of passive solar energy. In ancient Greece entire towns were designed so that all the homes could have optimal southern exposure to be cool in summer and warm in winter. As technology developed and our way of life came to be based on fossil fuels, we burn nonrenewable resources to power machines that make heat and cold. Unless, or until, we can orient our homes to the sun, we need to choose machines that use energy as efficiently as possible (see **HEATING SYSTEMS**).

In the vernacular architecture of the Middle East, buildings had air scoops to catch the cool evening breezes. Sustainable cooling systems would utilize the coolness of the northern exposure, shade trees, the temperature of the soil, and flowing water. I personally don't ever think about air conditioning—in the summer the fog comes into San Francisco Bay most afternoons to cool us off.

In addition to their energy use, air conditioners have another environmental concern—ozone-depleting CFCs and HCFCs (see Chapter 10). The most sustainable choice is to cool your house by natural means such as planting shade trees (the energy conservation organizations listed in the Directory of Organizations have more tips).

You can reduce your need for air-conditioning by minimizing the amount of heat that is created in your home. The three major sources of heat gain are heat that is conducted through your walls and ceilings from outside air, waste heat that is given off inside your house by lights and appliances, and solar heat gain from the sun shining through your windows. To minimize these heat sources, weatherize your home, use energy-efficient appliances and lights, and shade windows with vegetation or an interior window covering.

If you need an additional appliance to cool your home, see **AIR CONDITIONERS**.

Water Heating

Water heating consumes 15 to 30 percent of the energy used in the home. We all want hot water—for the most sustainable and efficient, see **WATER HEATERS**.

Appliances

Appliances account for about 30 percent of the energy use in the average American home. The most energy-efficient appliances currently on the mass market can reduce that figure by about one third, saving you 10 percent or more on your total energy bill.

Refrigerator-freezers take the most energy, followed by clothes washers, dishwashers, clothes dryers, cookers, and televisions. Gas appliances generally use up to 50 percent less energy than electric ones, and run on more-abundant and cleaner-burning natural gas or propane, rather than electricity generated by fuel from crude oil or nuclear power. If you do have electric appliances such as a dishwasher or clothes washer and dryer, run them at off-peak hours, like late at night. To meet the demand during peak hours, utilities often must use backup generators that are even less energy-efficient.

Although replacing an appliance before it breaks may not be the most efficient use of resources, sometimes it can be energy- and cost-effective to do so.

See **COOKING APPLIANCES, COOKWARE, DISHWASHERS, CLOTHES WASHERS, CLOTHES DRYERS, AND REFRIGERATORS AND FREEZERS**.

Light Bulbs

One of the icons of environmental products is the energy-efficient light bulb. Lighting accounts for only 6 percent of our total energy use nationwide, but 20 percent of all electricity produced. Light bulbs are one of the most inefficient of all products—a standard incandescent light bulb converts only 6 to 10 percent of the energy used to light (the rest ends up as heat), with so-called energy-saving incandescents being only about 1 to 5 percent better.

Proper lighting is a combination of natural light, overhead lights, and task lights. Different types of bulbs are suited to different applications; for the type that will best suit your needs, see **LIGHT BULBS**.

Automobiles

Since 63 percent of all the crude oil in America goes toward making gasoline for our automobiles, and driving cars contributes to so much air pollution, smog,

TYPICAL ENERGY CONSUMPTION OF VARIOUS HOUSEHOLD PRODUCTS

While the energy use of these products is, on a national level, relatively small, in individual households it is not uncommon to find that one or more of these products is using more energy than the refrigerator, water heater, or even the heating system. Added together, these small uses could account for a good percentage of your energy consumption.

APPLIANCE	TYPICAL ENERGY USE (KWH/YEAR)
Aquarium/terrarium	200 to 1,000
Auto block heater	150 to 800
Bottled water dispenser	200 to 400
Ceiling fan	10 to 150
Clock	17 to 50
Coffee maker	20 to 300
Computer	25 to 400
Crankcase heater	100 to 400
Dehumidifier	200 to 1,000
Electric blanket	75 to 200
Electric mower	5 to 50
Furnace fan	300 to 1,500
Garbage disposal	20 to 50
Grow lights and accessories	200 to 1,500
Humidifier	20 to 1,500
Instant hot water	100 to 400
Iron	20 to 150
Pipe and gutter heater	30 to 500
Pool pump	500 to 4,000
Spa/hot tub (electric)	1,500 to 4,000
Sump/sewage pump	20 to 200
Television (black & white)	10 to 100
Television (color)	75 to 1,000
Toaster/toaster oven	25 to 120
VCR	10 to 70
Ventilation fan	2 to 70
Waterbed heater	500 to 2,000
Well pump	200 to 800
Whole-house fan	20 to 500
Window fan	5 to 100

Source: *Consumer Guide to Home Energy Savings.*

and other environmental problems, perhaps the most important product to choose for energy-efficiency is our car (see **AUTOMOBILES**).

Renewable Energy

In order to be sustainable, ultimately—within our lifetimes, I believe—we as a society will have to switch from fossil fuel energy sources to renewable energy sources. Renewable energy sources with workable technologies available today are solar power, biomass (plant and animal waste used to make fuel), solar-derived hydrogen fuel, ecologically sized hydroelectric power, and wind power. Solar and wind-powered electricity in appropriate locations are already cost competitive and are becoming more widely known and available.

Studies for the U.S. Department of Energy say that within forty years the United States could get 57 percent to 70 percent of the total energy it uses now from sun, wind, water, geothermal, and biomass. In *State of the World 1991* World Watch Institute predicts that by 2030 photovoltaic cells (which convert sunlight directly into electricity) "will almost certainly be ubiquitous." Over the past twenty years the cost of photovoltaic energy has dropped from thirty dollars per kilowatt-hour to just thirty cents. By the end of the century, improvements in cell efficiency and increased demand are expected to bring the price down to ten cents per kilowatt-hour, and then it will be less than what I pay today to get my electricity from Pacific Gas & Electric.

Renewable energy sources are not completely environmentally harmless (though much less so than the by-products of fossil fuels or nuclear energy) and they are not unlimited. But properly managed and used, they can provide a fixed-rate flow forever. Combined with energy efficiency, we can learn to have the services we need or want within the amount of energy that is sustainably available. In addition to reducing the amount of fossil fuels used, energy efficiency is also a first step we can all take toward energy sustainability (see **ENERGY, RENEWABLE**).

WATER

It may seem unnecessary to conserve water when we see so much around us. Water is an abundant renewable resource of the earth. Through the many stages of the great, worldwide hydrologic cycle, water moves through the atmosphere as rain, snow, hail, dew, fog, and mist, and is stored within and on the surface of the earth in aquifers, springs, streams, rivers, lakes, oceans, and icebergs.

With all of this water all around us, why do we need to worry about running out of water?

The problem is not so much that there isn't any water, but rather that not all the water on this planet is usable by us. About 97.2 percent of the earth's water is salt water. The remaining 2.8 percent—fresh water—is used for the natural functions of the planet or is flood runoff and quickly returns to the sea.

About 2,376 trillion gallons a year can be captured for human use, which, according to an article in *Scientific American*, is enough to sustain 20 billion people if the water were evenly distributed. However, the total amount of water we use worldwide per year has risen since 1950 from about one tenth this amount to close to half this amount in 1980. At this rate of increase in both population and per-person water use, we will exceed our capacity within ten to twenty years. Even with the world's present population of 5 billion, if everyone used as much water per person as we currently do in the United States (580,000 gallons per person per year, about 1,600 gallons per day), we would already be using more water than is available worldwide!

Between global warming, poor agricultural practices, and more people using more water, less and less water is available for us for our daily needs. The loss of our forests particularly has played a major role in the loss of available water. Forests protect watersheds, protect soil from erosion, form windbreaks, and provide shade. Severe wind and rain erosion, the silting of rivers and irrigation systems, and the increasing severity of drought and floods around the world are the result of our loss of forest cover. When forests go, rivers often first flood and then dry up, since forests hold in moisture and recycle water back to the air through evaporation and transpiration.

Whether or not you need to be concerned about saving water is a very local issue. In an area such as Seattle, there is much local rainfall. Here in northern California, we just ended six years of drought, a longer drought than I've ever experienced. Because all our water comes from rainfall and runoff where I live, lack of rain means the reservoirs are low and we have to conserve. In 1992 many counties were forced to go on water rationing. Had we been more efficient about our use of water in the wet years, there would have been more available to us in the dry years.

Also in northern California we are faced with a growing population but only a fixed amount of water, which means that each person is allotted less. Our water departments are now promoting water conservation as a way of life, simply so they will be able to supply our residents without importing water from other areas.

Using water sustainably means using no more water than can be naturally replaced. Municipal tap water comes from one of two sources: surface water or an underground aquifer.

Surface water is renewable. It is produced by collecting the runoff from rain and snowmelt in a reservoir from an area known as a watershed. Often watersheds and reservoirs are open to the public for recreational purposes. If there is normal or heavy rainfall during the year, more water is available; if there is less rainfall, the amount of available water becomes less and less. To be sustainable, the amount of water taken from the reservoir on an annual basis can be no more than is replaced by rainfall or snowmelt during that year. If more water is used during successive years than is replaced, then the reservoirs eventually run dry. When there is a balance between the amount that is used and the amount that is replaced, then there is a continuous supply of water.

WATER-EFFICIENT FOODS

Here in America and worldwide, agriculture uses more water than people do—75 percent of all water is used for agriculture (25 percent for produce and 50 percent for livestock). Foods are not labeled with the amount of water it takes to grow or raise them, so here are some figures to help you create a "water-efficient" diet, if you choose.

GALLONS OF WATER NEEDED TO PRODUCE ONE SERVING OF:

Tomatoes	3
French fries	6
Lettuce	6
Sugar	8
Cola soft drink	10
Wheat bread	15
Apples	16
Oranges	22
Cantaloupe	51
Corn	61
Milk	65
Cherries	90
Margarine	92
Watermelon	100
Eggs (2)	136
Chicken	408
Pork	408
Hamburger	1,303
Steak	2,607
Typical breakfast	209
Typical lunch	1,427
Typical dinner	2,897
Daily total (for one person)	4,533

Source: *Los Angeles Department of Water and Power.*

One problem with maintaining a sustainable supply is that in many areas, surface waters originate in places far from the customer's tap. My tap water comes from a rain-fed reservoir about a mile away that I can hike up to and physically see how much water is there, but this is the exception rather than the rule. Across San Francisco Bay in Oakland, for example, the water comes from a reservoir in the Sierra Nevada mountains several hundred miles away, near Yosemite. And we have water shortages here because much of our water goes to Los Angeles where, by nature, there is practically none.

Underground aquifers supply other areas. Wells and springs bring up water from deep inside the earth. Water from aquifers is not renewable. Though they are constantly being replenished by water filtering through the earth's crust, aquifers cannot be refilled within our lifetimes with fresh, pure water.

Once the water has been collected, it must be purified, pumped, and transported to the taps—a process that utilizes resources, energy, and a considerable amount of toxic chemicals. Heating water at home for showers and washing also uses energy. So regardless of the local rainfall, it is environmentally prudent to conserve water as a precious resource.

WATER-EFFICIENT, WATER-SAVING

Like energy-saving and energy-efficient, water-efficient and water-saving are two similar terms used to describe a reduction in water use, but when referring to water, the words *efficient* and *saving* are used interchangeably. Both indicate that a product utilizes less water than other similar products to accomplish the same end result.

I have not found any federal or state regulations or guidelines regarding the use of terms that indicate water savings. The only related law I found was in California, which requires ultra-low-flush toilets in all new construction, and adoption of a water-efficient landscape ordinance by cities and counties to encourage businesses and developers to plant water-saving landscapes. Similar laws may be passed in other states in the future.

Water-Efficient and Water-Saving Products

A single person typically uses about thirty thousand gallons of water a year indoors. Toilets use more water in our homes than any other activity—about 30 percent. The amount of water used to take a shower comes in a close second to toilet flushing, depending on how long your shower is. Running the faucets accounts for only about 15 percent of household water usage, but the amount of water running out of an unrestricted faucet is enormous—up to fifteen gallons per minute.

Your local water district should have pamphlets for water-saving activities you can do at home (such as turning on the faucet to a lower force or not letting the hot water run constantly while you are washing dishes), so I won't go into listing them here. But there are a specific handful of products on the market that are worth purchasing, designed to make our water use more efficient so we get the same water performance for a smaller amount of water.

It takes no more than a few hours and a few dollars to purchase and install a low-flow shower head (see **SHOWER HEADS**) and faucet aerators (see **FAUCETS**). And for a couple hundred dollars, you should be able to install a low-flow toilet (see **TOILETS**). If you haven't already taken these water-saving measures, make it a Saturday morning project.

CHAPTER 9

Nontoxic

The idea of sustainability, by its very nature, is about the support of all life on earth. The way life sustains itself over time is through a continuous cycle of beginnings and endings—the seasons pass and come back again, flowers bloom and wilt and new blossoms spring from their seed, water evaporates and returns to the earth as rain. Even in the most seemingly catastrophic of natural disasters, nature is just clearing the way and putting the resources into the renewal of life.

In contrast, in our technological world we have created myriad substances which can destroy life in ways that make renewal impossible in the natural time frame. When we see products on the shelves in our local stores, we make an assumption that they are safe to use or they would not be available for sale. In fact, almost every kind of product on the market might have some component that could be harmful to health or the environment, and many products utilize toxic substances in their manufacture.

Only those products that do not create health risks or pollute the environment and that allow the regeneration of life can truly be called sustainable.

HUMAN TOXICITY

While there certainly have been instances of toxic products throughout history—the Mad Hatter was not just an *Alice in Wonderland* character; hat makers then regularly went mad as a result of occupational mercury poisoning—we live in a unique time, when potentially hazardous substances might be present in almost every type of product we purchase.

Toxicity of chemicals is of two types: acute and chronic.

Acute toxicity refers to poisoning as the result of a one-time exposure to a relatively large amount of a chemical. Acute toxicity is a concern for most consumer products that have warning labels, and the reason we have poison control centers. Every year 5 to 10 million household poisonings are reported as the result of accidental exposure to toxic products in the home. Some are fatal, and most of the victims are children. These poisonings are the result of accidental ingestion of common household products that, despite warning labels, are not kept out of children's reach.

Chronic toxicity refers to illness as the result of many repeated exposures to small amounts of a chemical over a long period of time, and this is what makes it so difficult to identify some toxics. We can easily see the effects when drain cleaner is spilled on someone's hand and the skin burns. The effects of chronic toxicity may not show up for years. Numerous common household products can cause cancer—not an immediate effect because carcinogenic substances take twenty years or more to act. Other household chemicals are mutagenic: They can change genetic material and lead to health problems. Still others are known to be teratogenic, and the high incidence of birth defects continues to remind us that not all household substances have been tested for this danger.

Toxic chemicals are made biologically by nature and in the laboratory by humans. Just because something is "natural" doesn't mean it's not toxic. Some of the most toxic substances in the world are in other plant and animal species, used for protection against predators. We don't generally encounter natural toxins in our everyday world, but we are exposed to many man-made substances in everyday products that have potentially toxic effects.

The basic problem is that nowadays most of our products are made from petrochemical derivatives of nonrenewable crude oil. Often these compounds are not otherwise found in nature, and our bodies have not developed the means to identify or assimilate them. These chemicals are used in nearly every industry and every type of consumer product.

In 1987 the United States produced almost 400 billion pounds of synthetic organic chemicals, more than four pounds of man-made substances for every person in this country each day. Worldwide, about seventy thousand synthetic chemicals are in use, with nearly one thousand new ones added every year. Some

HOUSEHOLD HAZARDOUS WASTE

One very important reason to buy nontoxic products is to eliminate your household hazardous waste. The prevalence of toxic products in the home has led most communities around the country to have household hazardous waste collection days. In many places it is illegal to throw certain toxic products in the trash. Here's the list of forbidden toxic products from my local household hazardous waste collection program. If you have these products around the house, dispose of them on a proper collection day or find out from your local solid waste department how to dispose of them.

There are nontoxic alternatives to almost all of these products.

House-Cleaning Supplies

Ammonia cleaners

Basin, tub, and tile cleaners

Chlorine bleach

Cleansers

Disinfectants

Drain openers

Furniture and floor polish

Lye

Metal polish

Oven cleaner

Rug cleaners

Laundry Supplies

Dry-cleaning solvent

Mothballs and flakes

Spot remover

Other Household Products

Aerosol cans

Butane lighters

Flea powder

Lighter fluid

Pet shampoo

Shoe dye and polish

Cosmetics

Cuticle remover

Depilatory cream

Hair-permanent solutions

Hair-straightener solutions

Nail polish

Nail polish remover

Medicines

Chemotherapy drugs

Liquid medicine

Mercury from a broken thermometer

Prescription medicine

Rubbing alcohol

Shampoo for lice

GARDEN SUPPLIES

Fungicides

Herbicides

Insecticides

Rat, mouse, and gopher poison

Snail and slug poison

Soil fumigants

Weed killer

AUTOMOTIVE SUPPLIES

Aluminum cleaner

Auto-body filler

Automatic transmission fluid

Brake fluid

Carburetor cleaner

Car wax

Chrome polish

Diesel fuel

Engine degreaser

Gasoline

Kerosene or lamp oil

Lubricating oil

Motor oil

BUILDING AND WOODWORKING SUPPLIES

Asbestos

Fluorescent lamps with ballasts and tubes

Glues and cements

Wood preservatives

HOBBY AND ART SUPPLIES

Acrylic paint

Artist's mediums, thinners, fixatives

Chemistry sets

Oil paint

Photographic chemicals/solutions

Resins, fiberglass, and epoxy

Rubber cement and thinner

PAINTING SUPPLIES

Latex paint

Model airplane paint

Oil paint

Paint stripper

Paint thinner, turpentine, mineral spirits

of these chemicals are considered safe for human use, but the vast majority have not been fully tested.

Next to nothing is currently known about the human toxic effects of almost 80 percent of the more than 48,000 chemicals listed by the EPA. Fewer than one thousand have been tested for immediate acute effects, and only about five hundred have been tested for their ability to cause long-term chronic health problems such as cancer, birth defects, and genetic changes. A National Research Council study found that complete health-hazard evaluations were available for only 10 percent of pesticides and 18 percent of drugs used in this country.

Almost no tests have been undertaken to evaluate the possible synergistic effects that occur when chemicals are combined in food, water, or air, or when chemicals interact with other chemicals in your body. The few studies that have been done indicate that such effects increase risks dramatically. Because scientists do not understand the ultimate effects of these chemicals, the government cannot begin to effectively regulate their use.

The average American home is literally filled with products made from these inadequately tested synthetic substances; we use more chemicals in our homes today than were found in a typical chemistry lab at the turn of the century. When professionals use chemicals in industrial settings, they are subject to strict health and safety codes, yet we use some of these same chemicals at home without guidance or restriction.

As further research is done scientists are finding that many household products we assumed to be safe are actually toxic to some degree or another. A multitude of common symptoms can be related to exposure to household toxics, such as headaches and depression. Insomnia, for example, is listed in toxicology books as a common symptom from exposure to the formaldehyde resin used on your bed sheets to keep them wrinkle-free.

Determining the toxicity of any product can be difficult because there are so many factors to consider. It's easy to identify some substances that are inherently unsafe, such as the bacteria that cause botulism, or benzene, which has known harmful effects, but tracking down the toxicity of other substances can be a much more complex process.

Toxicity of a substance is scientifically determined primarily through the use of animal studies in controlled laboratory experiments, although studies are sometimes made on human subjects if the effect of the substance is thought to be reversible. An experiment to find acute effects (known as the LD_{50}) determines the dose that causes the immediate death of 50 percent of the animals. Experiments on chronic effects might look for changes in blood chemistry, enzyme activity, tissue damage, and cancer induction over a period of several months, or even years. Some experiments require several generations of animals. Though animal rights activists may object, the government requires these animal tests by law to assess the potential hazards of compounds before they are released for use by the general population.

Epidemiological studies are also used. They show a statistical correlation between the occurrence of disease in a population and the factors suspected of causing that disease. Such a study often begins with a clinical observation, such as an unusually high frequency of cancer among those who smoke. These studies are valuable because they are based on the actual occurrence of real diseases in humans.

Whether or not a particular substance actually creates a toxic effect in you depends on:

- The quantity of the substance you are exposed to
- The strength of the substance (a small amount of one substance might be much more harmful than a large amount of another)
- The method of exposure (ingestion, inhalation, or skin absorption)—some substances are safe to inhale, but not to eat or rub on your skin; others are dangerous regardless of how you are exposed
- How frequently you are exposed—many substances have a cumulative effect in the body and do not cause harm until a certain concentration is reached over repeated exposure
- Your own individual tolerance for a substance.

After years of studying all the complexities of the toxicity of consumer products, the best advice I can give is that rather than trying to figure out if something that is inherently toxic is safe for you, just use as many nontoxic alternatives as you can. The more we stay away from toxics, the better off we will be, for our health and the environment.

ECOTOXICITY

Beyond the issue of human toxicity is the question of whether a substance is *ecotoxic*—that is, toxic to other species in the environment and to the sustainability of the functions of the earth. Every living thing has its own tolerance for toxics and the potential for harm to other species is sometimes greater and sometimes less than for us. We are much more sensitive to radiation, for example, than are most plants and animals, yet some pesticides are more dangerous to beneficial insects and fish than to humans.

One example of ecotoxicity occurred about fifteen years ago, when the pesticide Malathion was sprayed from helicopters over many residential areas in northern California to kill the Mediterranean fruit fly. The spraying showed no immediate human health effects, but later the deaths of thousands of fish and the loss of many beneficial insects were linked to the chemical.

The ecotoxicity of a substance is generally determined by evaluating the inherent toxicity of the substance, its persistence in the environment, and its tendency to bioaccumulate up the food chain.

Establishing the inherent toxicity of a substance has a lot to do with the organism that is receiving the exposure—a flower, tree, insect, fish, bird, or whale. We know a tremendous amount about toxic exposure to experimental laboratory animals, but beyond that, our knowledge is limited to the toxic effects on only a few species of birds and fish.

Toxic synthetic chemicals have a tendency to persist in the environment because their molecular structures do not break down under normal conditions. Once these artificial compounds are made, the environment does not have any way to decompose them and recombine them into other usable substances, and they become "pollution." PCBs (an oil additive), for example, were developed soon after the use of electricity became widespread, because electrical equipment needed oil that didn't break down. Now we find that PCBs are toxic and we cannot escape them because they persist in the environment. This is true for most other man-made substances as well.

Persistent substances that are not easily broken down accumulate in organisms low on the food chain, which in turn are eaten by predators higher up. As you go up the food chain, each higher predator has a greater and greater accumulation. Humans are at the top of the food chain and are thus one of the most bioaccumulative organisms. The result is that we all carry around high levels of chemicals in our fat that in the environment are at much lower levels, and pass this accumulation on to our offspring.

A major concern about toxics in the environment is that within the air, land, and especially water, toxic compounds can combine with each other to form other chemicals. Water is excellent for speeding up chemical reactions, making our waterways a kind of chemical reactor that can change the toxicity of many substances.

But even with all the money that might be spent on future environmental tests, we may still never be able to determine ecotoxicity accurately. Substances can bioaccumulate in one organism with no visible effects yet be fatally toxic to its predator. Chemicals can subtly interfere with plant and animal behavior or reduce the population of a species, affecting the whole predator-prey relationship of an ecosystem.

Obviously, we could initiate complex programs that include multispecies tests, ecological testing preserves, and mathematical ecosystem models to give us accurate data, but is the effort worth it? Perhaps the health and environmental costs of using toxic man-made chemicals have already exceeded their benefits.

Unfortunately, at present we, as consumers, cannot evaluate ecotoxicity on a product-by-product basis because we don't have adequate information on the toxics released during manufacture or use. And there is little research available on the ecotoxicity of various chemicals. For now, the best we can do for most substances is assume that if it is toxic to humans, it is probably also toxic to the environment, and that if a product contains a known toxic substance, its manufacture probably produces toxic waste.

ECOTOXICITY TEST FOR AQUATIC LIFE

My local California Department of Fish and Game (Region 3) has been testing the toxicity of common consumer products that end up in our waterways. They do a test called the LC_{50} (Lethal Concentration 50), which reveals the concentration in water of any substance that would kill half the aquatic organisms in ninety-six hours (this is actually a theoretical number—they find the concentration at which all die and at which none die, then calculate the concentration at which half die). Here are the results of some of the products they've tested:

PRODUCT	LC^{50}	
Household Bleach	4 ppm	(most toxic)
"Natural Solvent" Cleaners and Degreasers	31 ppm	
Nokomis All Purpose Cleaner Concentrate	36 ppm	
All Laundry Detergents	44 ppm	
Sunlight Dish Detergent	49 ppm	
Eagle One Car Wash and Wax, Super Concentrate	114 ppm	
Kelly Moore Premium Acry-Shield Paint	275 ppm	
Amway L.O.C. (Liquid Organic Cleaner)	315 ppm	
Fabergé Organic Shampoo	1,300 ppm	
Kelly Moore Latex Flat Wall Paint	1,650 ppm	
Hydrogen Peroxide (3% solution)	1,675 ppm	
Tone Bar Soap	10,000 ppm	
Hawaiian Punch Fruit Drink	27,500 ppm	(least toxic)

I find it interesting that laundry detergents are more toxic to fish than paint(!), even though to humans it is the other way around. But I'm not surprised that household bleach was the most toxic thing they tested, since this is the same hypochlorite used to kill all the bacteria in our municipal water supplies.

NONTOXIC

When we read the word *nontoxic* on a product label, our first impression is that the product is "not toxic." Scientifically, however, there is no experiment that can prove something is nontoxic. In an article in the October/November 1992 issue of *Garbage* magazine, toxicologist Dr. Alice Ottoboni, formerly of the California State Department of Public Health wrote, "It is not possible to prove a negative . . . Many of the questions asked about the effects of environmental chemicals . . . are unanswerable by science because science cannot conduct the experiments necessary . . . Toxicologists can answer all questions about what quantities of exposure would be harmful, but they cannot answer many questions about what quantities of exposure would be absolutely harmless."

With regard to "tolerances" set for substances such as food additives and pesticide residues, "acceptable daily intakes" for all sorts of environmental chemicals, and "threshold limit values" set for workplace exposures, Dr. Ottoboni says, "There is no absolute proof that these standards are totally protective for all people . . . Toxicologists are fully aware of the deficiency of such standards, but consider that they are protective for the great majority of people because they are set using large margins of safety."

Federal Regulations

There is no legal or regulatory definition of nontoxic that I have been able to find, though manufacturers tend to use nontoxic if a product does not meet the legal definition of toxic. According to Part 1500 of the 1960 Federal Hazardous Substances Act:

> "Toxic" shall apply to any substance (other than a radioactive substance) which has the capacity to produce personal injury or illness to man through ingestion, inhalation, or absorption through any body surface.
>
> "Highly toxic" means any substance which falls within any of the following categories:
>
> (A) Produces death within 14 days in half or more than half of a group of 10 or more laboratory white rats each weighing between 200 and 300 grams, at a single dose of 50 milligrams or less per kilogram of body weight, when orally administered; or
>
> (B) Produces death within 14 days in half or more than half of a group of 10 or more laboratory white rats each weighing between 200 and 300 grams, when inhaled continuously for a period of 1 hour or less at an atmospheric concentration of 200 parts per million by volume or less of gas or vapor or 2 milligrams per liter by volume or less of mist or

EPA CLASSIFICATION SYSTEM FOR CARCINOGENS

Group A Human Carcinogen	There is sufficient evidence from epidemiological studies to support a cause-effect relationship between the substance and cancer.
Group B Probable Human Carcinogen	B_1. There is sufficient evidence from animal studies and limited evidence from epidemiological studies.
	B_2. There is sufficient evidence from animal studies, but epidemiological data are inadequate or nonexistent.
	There is limited evidence from animal studies and no epidemiological data.
Group C Possible Human Carcinogen	Data is inadequate or completely lacking so no assessment can be made (does not mean it is not carcinogenic).
Group D Not Classifiable as to Human Carcinogenicity	Substance has tested negative in at least two EPA-defined animal cancer tests in different species and in adequate epidemiological and animal studies.
Group E Evidence of Noncarcinogenicity for Humans	Classification is based on available evidence and substances may prove to be carcinogenic under certain unknown conditions.

dust, provided such concentration is likely to be encountered by man when the substance is used in any reasonably foreseeable manner; or

(C) Produces death within 14 days in half or more than half of a group of 10 or more rabbits tested in a dosage of 200 milligrams or less per kilogram of body weight, when administered by continuous contact with the bare skin for 24 hours or less.

Later in the document, the Hazardous Substances Act notes that "highly toxic" could also refer to "a substance determined by the [Consumer Product

SIGNAL WORDS FOR TOXIC PRODUCTS

Category	Signal Word	Approximate Amount Needed to Kill an Average Person	Precautionary Statement: Oral, Inhalation, or Dermal Toxicity	Precautionary Statement: Skin and Eye Local Effects
I Highly toxic	DANGER POISON	A few drops to 1 teaspoon	Fatal if swallowed (inhaled or absorbed through skin). Do not breathe vapor (dust or spray mist). Do not get in eyes, on skin, or on clothing. (Front panel statement of treatment required.)	Corrosive, causes eye and skin damage (or skin irritation). Do not get in eyes, on skin, or on clothing. Wear goggles or face shield and rubber gloves when handling. Harmful or fatal if swallowed. (Approximate first-aid statement required.)
II Moderately toxic	WARNING	1 teaspoon to 1 ounce	May be fatal if swallowed (inhaled or absorbed through the skin). Do not breathe vapor (dust or spray mist). Do not get in eyes, on skin, or on clothing. (Appropriate first-aid statement required.)	Causes eye and skin irritation. Do not get in eyes, on skin, or on clothing. Harmful if swallowed. (Appropriate first-aid statement required.)

III Slightly toxic	CAUTION	More than 1 ounce	Harmful if swallowed (inhaled or absorbed through the skin). Avoid breathing vapors (dust or spray mist). Avoid contact with skin, eyes, or clothing. (Appropriate first-aid statements required.)	Avoid contact with skin, eyes, or clothing. In case of contact, immediately flush eyes or skin with plenty of water. Get medical attention if irritation persists.
IV Not toxic or nontoxic	(none required)		(none required)	(none required)

Safety Commission] to be highly toxic on the basis of human experience."

So with these definitions we could assume that a nontoxic substance is one that does not have the capacity to produce personal injury or illness to humans through ingestion, inhalation, or skin absorption. The scientific measure for toxicity depends on animal studies. If half or more than half the animals die, then the test substance is highly toxic; if less than half die, it's not. There we have it. Up to half of the test animals can die and the product can be called "nontoxic." This certainly isn't the same as the mere negation or absence of the capacity to produce personal injury or illness, nor does it mean that a nontoxic product is completely safe.

Based on these animal tests, the EPA has defined four categories of immediate acute toxicity that correspond to different dose levels received through ingestion, inhalation, and skin contact (see sidebar).

At one time, these signal words accurately indicated the dose required to cause a toxic effect, but because of poor labeling practices, these words now give only a general degree of danger.

To indicate long-term chronic hazards, the EPA has a classification scheme for cancer-causing agents (see sidebar), based on animal tests and epidemiological studies. Based on an assessment of the weight of evidence, these classifications are similar to those developed by the International Agency for Research on Cancer and the World Health Organization. Classification of a substance may change as new evidence, improved testing methods, or better analytical techniques become available, which may also affect regulations.

On product labels, however, you are likely to find only the warnings for acute effects possible during use or accidental exposure.

State Regulations

In 1986 California passed Proposition 65, the Safe Water and Toxic Enforcement Act, which requires direct public warning of significant exposures to chemicals that are known to the state to cause cancer or reproductive harm. Such exposures might occur through the ambient environment, the workplace, food, or consumer products.

At face value this law may seem to require a warning label on all products that apply, but in practice this hasn't happened. We now have warning labels on gasoline pumps and at the entrance to every establishment that serves alcoholic beverages, but not on consumer products. Signs posted in the front windows of every supermarket give an 800 number you can call for the warnings. Hardly convenient when you're walking down the cleaning products aisle.

Nontoxic Products

Almost every item in our homes can have toxic components, but to cover every possible toxic exposure requires a book in itself (and there are several good ones; see Recommended Reading).

The word *nontoxic*, however, is generally used only on those types of products that have significant known toxicity—such as **CLEANING PRODUCTS** and **PEST CONTROLS**—to indicate that the product is a safer alternative to a more toxic formulation. To identify products that do not contain toxic ingredients, the word *nontoxic* is frequently used on the labels.

It is often easier to determine that a product is toxic, and choosing nontoxic products can be a matter of looking for labels that indicate toxicity, rather than looking for the word *nontoxic*.

Ironically, toxic products are not required by law to completely list their ingredients. One way to find out what's in a product that does not list ingredients is to contact the manufacturer and ask for their Material Safety Data Sheet (MSDS) for the product. An MSDS lists the ingredients, the manufacturer, hazards to safety and health, and precautions to follow when using it. The Household Hazardous Waste Project has an "MSDS Fact Sheet" if you need help deciphering the form.

Sometimes the information on a MSDS can be very revealing. I once requested the MSDS on a well-known cleaning product that advertises itself as nontoxic, socially responsible, and otherwise committed to the environment, and found glycol ether as the main ingredient. After I couldn't find this in my chemical dictionary, I called and was told that glycol ether was a class of substances and the specific glycol ether used was butyl cellosolve. This I did find in my dictionary as ethylene glycol monobutyl ether, a chemical so toxic that a major chemical company decided to no longer manufacture it. Once they diluted it down to 2 percent (98 percent of the product's plastic bottle is full of water) the toxicity studies came out relatively harmless, but I had to ask myself if a product made from a diluted toxic chemical is really environmentally sustainable, or even ethically should be called nontoxic. Now in 1994, a front-page story in my local newspaper reported that this company is being sued for misrepresentation of environmental claims.

One limitation to MSDS sheets is that they only require reporting hazardous ingredients present in concentrations greater than 1 percent or .1 of 1 percent for carcinogens. Thus they are not a complete listing of ingredients, or even a complete listing of hazardous ingredients.

In addition to labels and MSDSs, we can also learn of toxic wastes put into the environment by companies through federal and local "right-to-know" programs. Title III of the Superfund Amendment and Reauthorization Act (SARA) requires 4 million businesses to provide hazardous chemical inventory data to a hierarchy of local, regional, and state public committees. This data includes information on hazardous chemical usage and storage, emergency response plans, and routine discharges of toxics that are permitted into air and water. Basic information on toxic discharges is compiled by the EPA and can be found in your local county library through the Toxics Release Inventory data base.

CHAPTER 10

No VOCs
CFC-Free
Ozone Friendly
Ozone Safe

One of the dichotomies of nature is that a single substance can be harmful to living things in one instance and vital to life in another. Ozone is such a substance.

Ozone is exceedingly rare. One molecule of ozone is composed of three atoms of oxygen, and fewer than ten molecules in every 1 million molecules of air are ozone. In our immediate atmosphere, ozone is a harmful air pollutant that irritates our lungs. In our planet's stratosphere, however, a layer of ozone helps regulate the earth's temperature and acts as the earth's natural sunscreen, protecting us by absorbing most of the ultraviolet radiation from the sun that would be hazardous to humans, animals, and plants.

To be sustainable, we must prevent the creation of ozone in the air we breathe and protect the existing ozone in the stratosphere.

ATMOSPHERIC OZONE

Ozone is formed in our immediate atmosphere by the reaction of two kinds of air pollutants—VOCs and nitrogen oxides—in the presence of sunlight. VOC is the acronym for a group of chemicals known as volatile organic compounds. Chemically, a VOC is any compound containing at least one atom of carbon, which includes every substance in organic chemistry. For regulatory purposes in the State of California, and in product certification programs, the definition of VOC excludes a whole list of chemicals that are regulated as toxic air pollutants or stratospheric ozone-depletors.

As smog, ozone irritates our noses, eyes, throats, and lungs, and reduces lung capacity, particularly in children, physically active people, and senior citizens. It also damages millions of dollars' worth of crops and buildings every year.

VOCs are released into the air from automobile exhaust and consumer products. While automobiles may seem to be a far greater source than your little can of hairspray, in California alone the annual emissions of 176 million pounds of VOCs from 30 million Californians using consumer products are the same as if 20 million cars were added, each driven an additional 10,000 miles that year.

NO VOCs, LOW VOCs

Products that claim to have no VOCs or low VOCs have been specially formulated to eliminate or reduce the amount of volatile organic compounds they contain.

State Regulations

The control of smog-producing VOCs is a regional problem which requires regulation appropriate and necessary for that area. California, New York, and New Jersey now have laws regulating VOCs in consumer products.

California law, for example, has set VOC emission limits for a number of products (see next page), which go into effect at different dates between now and the end of the decade. For some products lower emission limits are phased in over several stages.

No-VOC and Low-VOC Products

There are many examples of consumer products that contain VOCs—the list of those regulated by the state of California is only partial. In general products in aerosol form contain the most VOCs, followed by liquids, gels, and sticks, with solids containing little or no VOCs, depending on the product. If you have a choice, avoid aerosol products as much as possible.

Two common products that are particularly high in VOCs and already have their own regulations are **Paint** and lighter fluid (see **Barbecue**).

CONSUMER PRODUCTS CONTAINING VOCS THAT ARE REGULATED BY THE STATE OF CALIFORNIA

Whenever possible, choose an alternative to these products or a VOC-free brand:

Aerosol cooking sprays

Air fresheners (sprays and solids)

Antiperspirants and deodorants

Automotive cleaners:
- Brake cleaners
- Carburetor-choke cleaners
- Engine degreasers
- Windshield washer fluids

Bathroom and tile cleaners

Dusting aids (aerosols and others)

Fabric protectants

Floor polishes and waxes

Furniture polish (all forms except solid or paste)

General purpose cleaners

Glass cleaners

Hair products:
- Hairspray
- Mousse
- Styling gel

Household adhesives:
- Aerosol
- Construction and panel
- Contact
- General purpose

Insecticides:
- Crawling bug
- Flea and tick
- Flying bug
- Foggers
- Lawn and garden

Insect repellents (aerosol)

Laundry prewash (all forms)

Laundry starch products

Nail polish removers

Oven cleaners (aerosols, pump sprays, liquids)

Personal fragrance products

Shaving creams

THE STRATOSPHERIC OZONE LAYER

Nature creates ozone in only two places—in the air from lightning bolts and at the earth's outer envelope, the ozone layer. Ninety percent of all the ozone around our earth is in the stratosphere. Maintaining the earth's stratospheric ozone layer is perhaps the most basic requirement for sustainability, for without this protective shield anything else we do to live sustainably will be futile. While we only first discovered the hole in the ozone a few short years ago, the scientific evidence is now indisputable: Our ozone layer is being destroyed by man-made chemicals, and at a rate far faster than was originally predicted.

In 1974 researchers at the University of Michigan and the University of California–Irvine first discovered the link between ozone depletion and certain man-made chemicals. They predicted that chlorofluorocarbons (CFCs) would not disintegrate quickly in the lower atmosphere, but instead would rise into the stratosphere before breaking apart to form chlorine monoxide and other compounds. There the highly reactive chlorine would split the bonded ozone molecules. But they had no idea that ozone depletion would be so severe or concentrated in certain parts of the world, and could not foresee how extensive or immediate the damage would be.

A British team, which had been measuring ozone levels over Antarctica since 1956, first began to notice irregularities in the early 1980s, but they didn't know if the hole was real, or if their instruments were malfunctioning. By 1985 they felt confident enough to announce their discovery, but it wasn't until 1987 that NASA flights into Antarctica found unusually high concentrations of CFCs and made a cause-effect connection between the suspected ozone-depleting chemicals and actual destruction of the ozone layer.

A combination of factors makes the Antarctic particularly vulnerable. First, the polar vortex winds collect chemicals that waft in from the industrialized world. Then the superfrigid air of the long Antarctic night causes clouds of tiny ice crystals to form high up in the stratosphere where they aid breakdown of the chlorine compounds. And finally the sun strikes the cold chlorine and ozone and sets off a chain reaction of ozone destruction that opens the hole. As the constant sunlight heats up the air, the winds slow down and air from the rest of the world spills over the edge and fills up the hole, minutely thinning out the ozone layer around the globe.

Scientists have now also discovered the beginnings of a similar problem above the Arctic, much closer to populated areas. In 1991 scientists reported that ozone levels had declined 4 to 8 percent over the northern hemisphere in the last decade. A wave of studies released in 1991 indicated that the ozone layer was thinning twice as fast as previously expected over highly populated regions, and is thinner for longer and over a larger area than scientists had thought.

The threat of ozone depletion also exists during warmer seasons. In October 1991 scientists preparing an assessment of ozone damage for the United Nations Environment Programme revealed that, for the first time, ozone over midlatitude regions like North America and Europe had thinned during the spring and summer. Depletion during these warmer seasons is more dangerous because the hole allows the sun's rays to penetrate when they're at their strongest and when people are more often outdoors.

In February 1992 a team of NASA scientists discovered levels of chlorine compounds in the atmosphere above North America that were as much as 50 percent higher than had previously been recorded anywhere. If the polar vortex (a fluctuating mass of winter air) had remained large and cold enough to last through late February, then elevated chlorine levels could have opened a temporary ozone hole where a loss of ozone of up to 40 percent would have occurred over densely populated areas of Europe, Asia, and North America. While the vagaries of nature spared us then, the same possibility will exist for many springs to come.

Meanwhile the major ozone hole over Antarctica has been growing larger and persisting longer. In 1992 the World Meteorological Organization reported that the ozone hole over Antarctica covered a record area, stretching over 9 million square miles, about three times the size of the continental United States. The hole, about 25 percent larger than in past years, spread over the southern part of Tierra del Fuego, a populated island on the tip of South America. Researchers attributed the increased severity in part to weather, but primarily to higher levels of man-made chemicals. In the Northern Hemisphere stratospheric ozone declined sharply over parts of Western Europe, and areas of depleted ozone have been found as far south as the Caribbean.

A 10 percent reduction in ozone is likely to lead to about a 20 percent increase in harmful ultraviolet radiation. Increased ultraviolet radiation leads to increased skin cancer and aging and wrinkling of the skin. A 10 percent depletion of the ozone layer could result in almost 2 million additional skin cancers each year. It also appears that increased ultraviolet radiation damages the human immune system at much lower doses than are required to induce cancer.

Ozone depletion can also cause eye damage. The United Nations Environment Programme has estimated that a 1 percent reduction in ozone could cause 100,000 to 150,000 additional cases of blindness worldwide, and a 10 percent reduction could cause up to 1.75 million cases of blindness.

Both land and aquatic ecosystems would be affected as well. Of more than two hundred plants studied, mostly crops, 70 percent were found to be sensitive to ultraviolet radiation. Ultraviolet radiation can also penetrate sixty feet or more into the sea. Studies show that phytoplankton, one-celled microscopic organisms that float on the ocean's surface and on which the entire marine food chain depends, decrease their productivity by about 35 percent with a 25 percent reduction in ozone. Scientists in Antarctica have documented a 6 to 12 percent actual decline in phytoplankton since 1987, which they attribute to high levels of radiation that penetrate the drastically depleted ozone layer over the continent.

We are already seeing the effects of ozone depletion—they are not just theoretical. Skin cancer and cataracts are increasingly common in Australia, New Zealand, South Africa, and Patagonia. In Queensland, in northeastern Australia, more than 75 percent of those over age sixty-five now have some form of skin cancer and children are required to wear large hats and neck scarves to and from school to protect them from ultraviolet radiation. In Patagonia blind rabbits and fish have been caught. Parents in Punta Arena, Chile's southernmost city, keep their children indoors between ten A.M. and three P.M. The Australian government issues alerts—like Los Angeles smog alerts—when especially high UV levels are expected, and public service campaigns warn of the dangers of sunbathing like our American ads warn against the dangers of drugs.

Ozone-Depleting Chemicals

Ozone-depleting chemical compounds contain molecules of chlorine, fluorine, or bromine. Title VI of the United States Clean Air Act Amendments of 1990 has defined ozone-depleting substances as belonging to Class I (most harmful) and Class II (less harmful).

Class I chemicals are the worst known stratospheric-ozone destroyers. Currently listed as Class I chemicals in Title VI are chlorofluorocarbons (CFCs), halons, carbon tetrachloride and 1,1,1-trichloroethane (methyl chloroform). These chlorine-based chemicals account for about 80 percent of ozone depletion.

CFCs have gotten the most publicity of all the ozone depletors. They are compounds comprised of carbon, fluorine, chlorine, and hydrogen. When CFCs were first developed in the 1930s by Du Pont, they were thought to be wonder chemicals because they were useful and nontoxic to humans. Only many decades later did we begin to learn how CFCs harm the environment.

Because they are lighter than air, CFCs rise through the atmosphere until they hit the ozone layer. When the reactive chlorine atoms of disintegrating CFCs collide with ozone (which is also unstable), a chlorine atom from the CFC steals an oxygen atom from the ozone molecule. Created in its place is a useless compound of one chlorine and one oxygen atom (chlorine monoxide) and a two-atom molecule of oxygen.

The oxygen atom of the chlorine monoxide molecule then starts searching for something to react with and bonds with the nearest oxygen atom, freeing the chlorine atom to find a new mate. By this constant change of dance partners, so to speak, a single chlorine atom can destroy 100,000 ozone molecules. Since this is happening a lot faster than nature can replenish the ozone layer, sending a continuous stream of CFCs into our stratosphere is not a sustainable activity.

Twenty million tons of CFCs having who knows how many chlorine atoms have already been released into the air. Those released today won't reach the stratosphere for ten to fifteen years. If we completely stop using ozone-depleting chemicals today, it will still take until the end of the next century to heal the ozone layer.

Halons are compounds of bromine and fluorine. There are three specific halons listed: halon 1211 (bromochlorodifluoromethane), halon 1301 (bromotrifluoromethane), and halon 2402 (dibromotetrafluoroethane). Carbon tetrachloride and 1,1,1-trichloroethane (methyl chloroform) both contain chlorine atoms. All these chemicals react with ozone similarly to CFCs.

Class II chemicals cause less damage to the ozone layer, but still pose a threat. The only Class II chemicals currently listed in Title VI are the hydrochlorofluorocarbons (HCFCs or HFCs), which are replacing the more ozone-depleting CFCs in many products. HCFCs have one-twentieth of the chlorine and more hydrogen, so they break down more easily in the atmosphere's lower levels. But organizations such as the Environmental Defense Fund and the Institute for Energy and Environmental Research oppose the use of HCFCs, saying that much of their chlorine still reaches the ozone layer.

Here in the United States, the manufacture and use of CFCs was prohibited for use in household aerosols in 1978, shortly after their ozone-depleting qualities were discovered. They are still allowed, however, in many other products (see below). By enacting the Clean Air Act Amendments of 1990, the United States became the first nation to actually legislate the complete phase-out of CFCs.

Once the ozone hole was confirmed, other countries also recognized the need to stop using ozone-depleting substances. In 1987 the Montreal Protocol (overseen by the United Nations Environment Programme) was signed by twenty-four countries agreeing to a 50 percent reduction in ozone-depleting chemicals by 1999. After scientific evidence subsequently suggested that the ozone shield was depleting more rapidly than had been previously thought, in 1990 ninety-three nations adopted a deadline phasing out the major chemicals altogether by the year 2000. In late 1992 these same countries agreed to move up the deadline, phasing out chlorofluorocarbons and carbon tetrachloride by January 1, 1996, eliminate halons by 1994, and ban methyl chloroform by 1996.

CFC-FREE, OZONE FRIENDLY, OZONE SAFE

CFC-free specifically means that the product does not contain or use chlorofluorocarbons. Ozone friendly, ozone safe, and similar terms mean that the product does not contain or use in manufacture CFCs or other chemicals known to destroy the ozone layer. While there are many products that are safe for the ozone because they never have and never will contain ozone-depleting substances, these terms are used to distinguish one product from another in product categories where ozone-depleting chemicals are commonly used.

Federal Guidelines and Warning Labels

There are no federal regulations regarding the use of buzzwords describing ozone safety; however, the Federal Trade Commission gives guidelines for use of the terms *ozone friendly* and *ozone safe*.

PROTECTING OURSELVES

While most of the United States is not yet in the worst zone of danger, the 4 to 8 percent ozone thinning that has occurred over the last decade makes it not too early to start protecting our health. UV rays cause harm after repeated exposure over long periods of time—not in one afternoon at the beach.

You can significantly reduce the risk of getting skin cancer or cataracts by taking the following precautions:

- Wear protective clothing when you are out in the sun for prolonged periods. Choose fabrics that have a tight weave and a wide-brimmed hat that covers your face, ears, and neck.
- In the summer when temperatures demand wearing clothing that leaves skin exposed, use a sunscreen with a skin protection factor (SPF) of at least 15.
- Stay indoors or in the shade between ten A.M. and three P.M.
- Wear sunglasses when outdoors in bright sunlight. Look for ones that are treated to absorb UV radiation.

They say "A claim that a product does not harm the ozone layer is deceptive if the product contains an ozone-depleting substance." An "ozone-depleting substance" includes all the chemicals listed in Class I and Class II of the Clean Air Act; therefore, any product labeled "ozone friendly" or "ozone safe" should not contain CFCs, halons, carbon tetrachloride, 1,1,1-trichloroethane (methyl chloroform), or HCFCs.

In addition, if a product makes an unqualified claim that it "Contains no CFCs" or is "CFC-free," it should also not contain HCFCs, since these claims may imply to reasonable consumers that the product will not harm the ozone layer at all.

If a product is labeled "This product is 95 percent less damaging to the ozone layer than past formulations that contained CFCs," the product may contain HCFCs if the manufacturer can substantiate that this substitution will actually result in 95 percent less ozone depletion.

Since May 15, 1993, warning labels have been required by the EPA, in accordance with the 1990 Clean Air Act, on products that contain or are manufactured with Class I substances. Any product that uses Class I chemicals must have a label conspicuously placed on the product that says it contains or is made with "a substance that harms public health and environment by destroying ozone in the upper atmosphere." The EPA hopes that the warning labels will give manufacturers an incentive to make and sell products that do not use Class I ozone-depletors.

The EPA believes that substitutes can eventually be found for all products currently using Class I chemicals, and has imposed a ban on their production by the end of 1995. Some will be replaced by the Class II HCFCs, but as soon as alternatives are available for HCFCs, there will be warning labels to discourage the use of these as well.

State Regulations

The term *ozone friendly* (or any similar term) is legally defined in the states of California and Indiana. Both agree that an ozone-friendly product is one that is made from chemicals or materials that will not, as a result of the use or production of the product, migrate to the stratosphere and cause unnatural and accelerated deterioration of the ozone layer.

CFC-Free, Ozone-Friendly, and Ozone-Safe Products

According to the United Nations Environment Programme, there has been a 40 percent drop in CFC consumption since 1986, a reduction beyond that required by the Clean Air Act and the Montreal Protocol. While this is an immense improvement, we still have a long way to go to get to zero ozone-depleting substances.

Businesses in industrialized countries are working to phase out ozone-depletors even ahead of the schedule in the Montreal Protocol. There are still many products, however, that use ozone-depleting substances, and we don't need to wait for the industry changes or government bans to take effect in order to reduce our own use of ozone-depletors.

One of the greatest releases of CFCs (50 to 60 percent) occurs when coolants escape from **REFRIGERATORS AND FREEZERS** when they leak, are repaired, or are junked. The common refrigerant Freon is a CFC. Federal law now requires that all CFCs be recycled when removed from refrigerators (both residential and commercial) and special equipment must be used to prevent their escape.

Another major release of CFCs is from home **AIR CONDITIONERS**. Check to make sure if your air conditioner uses CFCs or the less harmful HCFCs, which are already commonly used. The same federal law on CFC recycling applies to home air conditioners, too.

Right up there with refrigerators are air conditioners in **AUTOMOBILES**. Compared to ounces of HCFCs in a home air conditioner, a car air conditioner uses pounds of CFCs. Air-conditioned cars and trucks use 25 percent of CFCs and cause more than 16 percent of ozone destruction.

Even though household aerosol spray cans do not contain Class I CFCs, many do contain Class I methyl chloroform (MCF) or Class II HCFCs (see sidebar). So don't make an assumption that the spray isn't causing some harm to the ozone layer. Whenever possible, choose products in a nonspray form or in a pump spray bottle. CFCs can also be found in foamed plastics (including cushions, carpet padding, foam insulation, and food containers).

Carbon tetrachloride, a known carcinogen in addition to being a ozone-depletor is used as an agricultural fumigant, a solvent, and for dry cleaning. While not permitted in products for home use, it is still used in industry.

Methyl chloroform (1,1,1-trichloroethane) is used in more than 150 household products (see sidebar for the most common). Consumer use accounts for about 54 million pounds of methyl chloroform yearly and alternatives are available for most, if not all, of these items.

Virtually any product that contains any kind of chlorine has the potential ability to harm the ozone layer. Those chemicals regulated by the Clean Air Act are only the worst offenders. Many CLEANING PRODUCTS contain chlorine, such as household BLEACH and SCOURING POWDER. Many PEST CONTROLS also contain chlorine—avoid those that have *chloro* in their name such as pentachlorophenol. Another potentially large source of ozone-depleting chemicals may be the chlorine used in sewage and water treatment, swimming pools, spas, and hot tubs.

Volatile bromine compounds are released from leaded gasoline, oil and gas recovery processes, and agricultural use.

The only product listed by the EPA as containing CFC-depleting halons is FIRE EXTINGUISHERS.

PRODUCTS THAT CONTAIN OZONE-DEPLETING SUBSTANCES

The EPA has identified the following products that contain Class I and Class II ozone-depleting substances (I've given only the major household products for which we can choose alternatives here—for commercial and industrial products see the *Federal Register*, Vol. 57, No. 86, May 4, 1992). All of these products use ozone depletors as part of their aerosol or pressurized containers.

PRODUCT	CLASS I	CLASS II
Pesticides		
Carpet flea spray	MCF	
Flying insect killers	MCF	
Foggers, indoor and outdoor	MCF	
House and garden insecticides	MCF	
Automotive products		
Brake cleaner	MCF	
Carburetor/choke cleaner	MCF	
Emergency tire inflator	CFC-12, MCF	HCFC-22
Engine cleaner	MCF	
Leak sealer	MCF	
Spray lubricants	MCF	
Tire cleaners	MCF	
Household products		
Artificial snow/glass frosting	MCF	
Drain cleaners/pressurized blowers	CFC-12	
Fabric protector	MCF	
Kitchen fire extinguishers	MCF	
Leather/suede cleaner/protector/conditioner	MCF	
Paint/graffiti remover	MCF	HCFC-22

Party confetti (plastic party streamers)	CFC	
Party horns	CFC-12	
Spot remover	MCF	
Spray shoe polish	MCF	
Water repellent	MCF	
Wood cleaners and protectors	MCF	
Cosmetics		
Deodorant		HCFC-22
Hair color spray	MCF	
Hair conditioner	MCF	
Hairspray	MCF	
Nail polish dryers		HCFC-22
Perfume		HCFC-22
Strong-hold hairspray	MCF	HCFC-22
Art supplies		
Fixative spray	MCF	
Spray adhesive	MCF	
Medical		
Bronchial inhalant medicines	CFC-11, 12, 114	
Contraceptive foam	CFC-12, 114	

CHAPTER 11

Biodegradable
Degradable
Photodegradable
Compostable

At the end of the cycle for all products, they must ultimately go back to the earth. Disposal can be sustainable and serve to restore the fertility and productivity of the land and the abundance of aquatic life, or it can litter, pollute, and destroy the ecosystems on which we depend.

Sustainable disposal of any product requires that its waste return to the earth and biodegrade. Nature biodegrades everything it makes back into basic building blocks, so that new living things can be made from the old. Materials biodegrade through the action of microorganisms in water and soil, which feed on the materials and cause them to decay. Every resource made by nature returns to nature—plants and animals biodegrade, even raw crude oil will degrade when exposed to water, air, and the necessary salts. Nature has perfected this system—we just need to learn how to participate in it.

By the time many resources are turned into products, they have been altered by industry in such a way that they are unrecognizable to the microorganisms and enzymes that return natural materials to their basic building blocks. Crude oil, for example, will biodegrade in its natural state, but once it is turned into plastic, it becomes an unsustainable pollution problem. Instead of returning to the cycle of life, these products simply pollute and litter our land, air, and water.

BIODEGRADABLE, DEGRADABLE, PHOTODEGRADABLE

Of all the environmental buzzwords, biodegradable has perhaps been the most misused and is perhaps the most difficult to qualify and understand. Because in the past there have been no guidelines or regulations, many products have called themselves biodegradable without any real justification. Unfortunately, the word *biodegradable* has frequently been applied to products that generally aren't (detergents or plastics) and almost never used for products that really are (soap or paper).

Logic tells us that a biodegradable product has the ability to break down, safely and relatively quickly, by biological means, into the raw materials of nature and disappear into the environment. These products can be solids biodegrading into the soil (which we also refer to as "composting"), or liquids biodegrading into water.

Technically, "degradable" means having the ability to break down, regardless of the means. However with regard to product labeling it is generally used simply as a shorter form of biodegradable.

"Photodegradable" means that the product will break down when exposed to sunlight.

A leaf is a perfect example of a biodegradable product: It is made in the spring, used by the plant for photosynthesis in the summer, drops to the ground in autumn, and is assimilated into the soil to nourish the plant for the next season.

The basic concept seems straightforward enough; however, there are several factors to consider in determining the biodegradability of a product or material.

The first is the question of the inherent biodegradability of the material. Any material that comes from nature will return to nature, as long as it is still in a relatively natural form. Therefore any plant-based, animal-based, or natural mineral-based product has the capability of biodegrading, but products made from man-made petrochemical compounds generally do not. When a man-made compound is formulated in a laboratory, combinations of elements are made that do not exist in nature, and there are no corresponding microorganisms to break them down.

The next issue is how long it takes for the material to actually break down. In nature different materials biodegrade at different rates. A leaf takes approximately a year to become part of the forest floor. An iron shovel, on the other hand, can take years to rust away to nothing, and a large tree can take decades to completely break down. Common sense tells us that any material will ultimately biodegrade, even if it takes centuries.

So what is the proper rate for a material to be biodegradable? It really de-

RATE OF BIODEGRADABILITY FOR COMMON PRODUCTS

Here's how long it takes for some of our commonly used products to biodegrade when they are scattered about as litter:

Cotton rags	1 to 5 months
Paper	2 to 5 months
Rope	3 to 14 months
Orange peels	6 months
Wool socks	1 to 5 years
Cigarette butts	1 to 12 years
Plastic coated paper milk cartons	5 years
Plastic bags	10 to 20 years
Leather shoes	25 to 40 years
Nylon fabric	30 to 40 years
Tin cans	50 to 100 years
Aluminum cans	80 to 100 years
Plastic 6-pack holder rings	450 years
Glass bottles	1 million years
Plastic bottles	Forever

pends on the material itself. Our leaf example suggests that the proper rate is that which is appropriate to the ecosystem. A liquid going into a waterway should biodegrade fairly quickly, whereas there's no harm done if it takes a while for a newspaper to break down. Plastics, on the other hand, will not biodegrade in anyone's lifetime, and certainly will never break back down into the petroleum from which they were made.

And then there is the question of what exactly does the product or material break down into, and are there any toxic substances formed along the way or as the end result. In his book *The Closing Circle*, ecologist Barry Commoner gives the example of the benzene unit in synthetic detergents being converted as it biodegrades into phenol (carbolic acid), a substance toxic to fish. To be truly biodegradable, a substance or material should break down into carbon dioxide (a nutrient for plants), water, and naturally occurring minerals that do not cause harm to the ecosystem (salt or baking soda, for example, are already in their natural mineral state and do not need to biodegrade).

And the characteristics of the environment that the substance or material is in can also affect its ability to biodegrade. Detergents, for example, might break down in a natural freshwater "aerobic" (having oxygen) environment, but not in an "anaerobic" (lacking oxygen) environment such as sewage treatment plant digestors, or natural ecosystems such as swamps, flooded soils, or surface water sediments.

Many products that are inherently biodegradable in soil—such as tree trimmings, food wastes, and paper—will not biodegrade when we place them in landfills because the artificial landfill environment lacks the light, water, and bacterial activity required for the decay process to begin. The Garbage Project, an anthropological study of our waste conducted by a group at the University of Arizona, has unearthed hot dogs, corncobs, and grapes that were twenty-five years old and still recognizable, as well as newspapers dating back to 1952 that were still easily readable. When we don't provide the conditions needed for biodegradable materials to naturally biodegrade, we end up with major garbage problems.

Once it is determined that a substance or material will actually biodegrade under particular conditions, then there is the problem of actually disposing of the product in those conditions and in an amount that can be sustained by the ecosystem that is receiving it. The sustainable rate of biodegradation is that amount which a given ecosystem can absorb as a nutrient, and if necessary, render harmless.

Soap, for example, is a natural, organic product that is inherently biodegradable. The soapy gray water from our shower and washing machine runs out into our garden and we have a lush growth of plants that attest to the biodegradability of our soaps (see SOAP, SHAMPOO, and LAUNDRY SOAPS AND DETERGENTS). However, if that same soap went down a sewage line that fed into the San Francisco Bay along with the soap used by the other million or more residents of the counties that surround the bay, we might have waves of soapsuds on our beaches, simply because we would be putting more soap into the bay than it has microorganisms to biodegrade. But in our backyard these particular soap products actually do biodegrade very quickly.

Oil spills are devastating not because oil doesn't biodegrade, but rather because the amount of oil is much greater than the number of microorganisms available to degrade it. It has been estimated that it will take fifty years for the oil spilled in 1989 by the *Exxon Valdez* to degrade. Lakes and streams have become polluted because the amount of sewage dumped into them has been overwhelming. As much as we need to consider the biodegradability of the product, we also need to consider the capacity of the system the biodegradable substance or material is being placed into.

Those who have attempted to define biodegradability for product labels run into the same dilemma encountered when defining recyclability: Should a product be called biodegradable if it inherently has the ability to biodegrade, or should it be called biodegradable only if it also is commonly disposed of in a way in which it really will biodegrade? For example, should a paper grocery bag be labeled biodegradable? It will biodegrade if placed in nature; however, it won't biodegrade in a landfill because the conditions aren't right.

I believe consumers should be informed if a product has the ability to biodegrade so we can then dispose of it in a way that allows biodegradability. But regulators think it is misleading to suggest that a biodegradable product that is likely to end up in a landfill be labeled as biodegradable. Because degradability offers little or no environmental benefit for products that will be disposed of in landfills or incinerators, environmental claims relating to disposability should not be made unless a biodegradable disposal option is currently available to consumers in the area in which the product is sold and the product complies with the requirements of the relevant waste disposal programs. Any product promoted as degradable should clearly and prominently disclose that the product is not designed to degrade in landfills. So we have to look for potentially biodegradable products ourselves, and dispose of them properly.

One other thing to watch out for on labels is the statement that the "active ingredients are biodegradable." If a product contains active ingredients listed as such, then there are sure to be some inactive ingredients that are not on the label. If only the active ingredients are biodegradable, then the active ingredients are not, and this would not be a truly biodegradable product. If a product is labeled biodegradable, then every ingredient should properly biodegrade.

Federal Regulations and Guidelines

To my knowledge, there are no federal regulations requiring products of any kind to be biodegradable, but there are definitions of biodegradability, test procedures, and guidelines for product labeling.

Environmental Protection Agency

In the Federal Toxic Substances Control Act (TSCA, popularly pronounced "tosca"), there are sixty-nine pages describing test methods defined by the EPA for determining the biodegradability of substances in water and soil. (CFR 40, Subpart D, 796.3100–3400.)

TSCA defines "ultimate biodegradability" as "the breakdown of an organic compound to CO_2 [carbon dioxide], water, the oxides or mineral salts of other elements and/or products associated with normal metabolic processes of microorganisms." In this case, the word "organic" means those compounds, whether natural or synthetic, that contain carbon.

Federal Trade Commission

The FTC gives guidelines for the use of the terms *biodegradable, degradable,* and *photodegradable.*

They say "An unqualified claim that a product or package is degradable, biodegradable, or photodegradable should be substantiated by competent and reliable scientific evidence that the entire product or package will completely break down and return to nature, *i.e.* decompose into elements found in nature within a reasonably short period of time after customary disposal."

In addition "Claims of degradability, biodegradability, or photodegradability should be qualified to the extent necessary to avoid consumer deception about: (a) the product or package's ability to degrade in the environment where it is customarily disposed; and (b) the rate and extent of degradation."

They give three examples.

A trash bag is marketed as degradable (with no qualification or other disclosure) after the marketer used soil burial tests to show that the product will decompose in the presence of water and oxygen. But because trash bags are customarily disposed of in incineration facilities or at sanitary landfills that are managed in a way that inhibits degradation by minimizing moisture and oxygen, the claim is deceptive. Degradation will be irrelevant for the trash bags that are incinerated and soil burial tests do not give any indication of how the bags will degrade in the often dry, anaerobic conditions of a landfill.

I agree that this example is deceptive. In my opinion, the most informative way to label this product would be "degradable in soil, with the presence of water and oxygen, but not in a landfill." With this label, we would know that the product had the ability to biodegrade and under what conditions. But because the FTC makes an assumption (albeit correct) that most, if not all, of these bags go to the landfill, such a product would simply not say it is degradable, rather than instructing the consumer as to a possible degradable disposal.

Their second example is of a commercial agricultural plastic mulch film advertised as photodegradable and qualified with the phrase "Will break down into small pieces if left uncovered in sunlight." Because the claim is supported by competent and reliable scientific evidence that the product will break down in a reasonably short period of time after being exposed to sunlight and into sufficiently small species to become part of the soil, the qualified claim is not deceptive.

And finally a soap or shampoo product is advertised as biodegradable, with no qualification or other disclosure. If the manufacturer has competent and reliable scientific evidence demonstrating that the product, which is customarily disposed of in sewage systems, will break down and decompose into elements found in nature in a short period of time, the claim would not be deceptive.

State Regulations

Biodegradable is defined in California and Indiana as a material that has the proven capability to decompose in the most common environment in which it is disposed within one year through natural biological processes into nontoxic carbonaceous soil, water, or carbon dioxide.

Photodegradable is also defined in California and Indiana. To be labeled as such, material must have the proven capability to decompose in the most common environment in which it is disposed within one year through physical processes, such as exposure to heat and light, into nontoxic carbonaceous soil, water, or carbon dioxide.

Certification Organizations

Scientific Certification Systems has developed standards for certifying biodegradability claims for soaps, detergents, and cleansers. They are designed to verify that products degrade safely and efficiently (even under worst-case circumstances), and do not build up to harmful concentrations in the environment. SCS considers how long it takes for the product to break down, the nature of the substances the product breaks down into, and the impact on the environment.

There are minimum requirements for certification, which apply to every component of a formulation. Biodegradable means that the entire product biodegrades, not just a single ingredient.

To simplify some very complex standards, SCS is basically concerned that a product biodegrade completely under both aerobic and anaerobic conditions into carbon dioxide, water, and minerals. They will not certify any product that breaks down into secondary products that are toxic to the environment, or one which contains phosphates. Certified products carry the statement "Biodegradable Product—Breaks down into carbon dioxide, basic minerals, and water."

Green Seal has proposed standards for cleaning products that are primarily concerned with biodegradability and toxicity. They too require that a product be biodegradable under aerobic and anaerobic conditions, and not be toxic to aquatic life. In addition, the product must not be toxic as defined by the Consumer Product Safety Commission (see Chapter 9) and must not contain phosphates, EDTA, or a list of heavy metals beyond certain levels. They also require that cleaners be packaged in plastic bottles or cardboard boxes with recycled content.

Biodegradable Products

There are many products that claim to be biodegradable, but may not be according to the definitions outlined in this chapter. I wish I could give you a tip and say here's a simple way to tell if a product is really biodegradable or not, but it's not that easy.

If you want to verify the biodegradability claim of a product by contacting the manufacturer, here's what to ask:

- How do you define *biodegradable*?
- What are the ingredients?
- Are all the ingredients inherently biodegradable?
- Will the product degrade in aerobic and anaerobic conditions (or in a landfill, or in a compost pile)?
- How long does it take for the product (or each ingredient) to biodegrade?
- What does the product (or each ingredient) break down into?
- Is the product or the substances that it breaks down into harmful to the environment in which it breaks down?
- Do you have a laboratory test or independent certification that verifies biodegradability?

These questions may result in helpful information, a barrage of unreadable test results, or no information at all.

Virtually any product made from plants, animals, or minerals (in their unaltered, natural state) is biodegradable under the proper conditions, including paper, natural fiber cloth, food, yard wastes, and soap. But in the marketplace, the word *biodegradable* has only been used on a few types of products, primarily cleaning products and plastics.

I have not done a complete survey of the market to verify the biodegradability claims of all cleaning products, so for now the only ones I can recommend with any certainty are those certified by Scientific Certification Systems (see **Basin, Tub and Tile Cleaners, Cleaning Products**, and **Drain Cleaners**). This does not mean, however, that these are the only biodegradable products on the market—they are simply ones that have paid for and passed the certification tests. Your own inquiry of local manufacturers will probably turn up other appropriate biodegradable products for your own cleaning needs, and, of course, you can always make your own from natural mineral baking soda and biodegradable organic lemon juice and vinegar.

Photodegradable plastics are supposed to disintegrate into small pieces when exposed to sunlight (manufacturers add a sun-sensitive component to the plastic to trigger degradation). Biodegradable plastic is intended to break up when exposed to microorganisms (a natural ingredient such as cornstarch or vegetable oil is added to achieve this result).

Garbage bags are the prime example of so-called degradable plastics and almost every major manufacturer now sells a degradable version. But sustainability requires that degradable materials break down completely by natural processes so that the basic building blocks can be used again by nature to make a new life form. Plastics made from petrochemicals are not a product of nature and cannot be broken down by natural processes. Therefore, despite how small the pieces of plastic may become, they are not and cannot be biodegradable.

COMPOSTABLE

"Compostable" is almost synonymous with "biodegradable," except it is limited to solid materials and does not refer to liquids.

Compost is a mixture of decaying organic substances, such as dead leaves, etc., used for fertilizing land. A product that is compostable is one that can be placed into a composition of decaying biodegradable materials, and which will eventually turn into a nutrient-rich material. Composting occurs in nature every day as fallen leaves and tree limbs biodegrade into the forest floor. The EPA considers composting a form of recycling because it turns resources into a usable product.

Compost piles have been used by farmers and gardeners for generations. Food, leaves, grass clippings, garden wastes, and tree trimmings (which amount to between 50 and 70 percent of waste in this country) can all go into the compost pile, where hungry microorganisms eat the waste to produce carbon dioxide, water, and humus. The resulting compost is an excellent natural fertilizer proven by organic gardeners and farmers to restore soil fertility, control weeds, retain ground moisture, and reduce soil erosion.

While backyard compost piles are well known, the newest application of composting is municipal composting, which works on the same natural principles, but is done on a much larger scale. More than 2,200 communities already compost their leaves, grass, and yard trimmings. Approximately fifty-five additional communities compost or are about to compost all their organic trash at well-sited, professionally managed, composting facilities.

Municipal composting requires minimal time, effort, and labor, since most of the work is done by the microorganisms. Communities can also use or sell the resulting compost for agricultural and horticultural use, or to restore depleted land. Unlike landfills, a composting site can be continually reused without ever reaching capacity.

As with the term *biodegradable*, regulators recommend that the term *compostable* not be used unless the product is currently composted in a significant amount in the area where it is sold. Without the ability to actually compost the product, a claim is considered to be meaningless and thus deceptive. Regulators recommend that any product promoted as compostable should clearly and prominently disclose that the product is not designed to degrade in landfills.

Federal Guidelines

There are no federal regulations regarding the use of the term *compostable*, but the Federal Trade Commission does give guidelines.

They say "An unqualified claim that a product or package is compostable should be substantiated by competent and reliable scientific evidence that all the materials in the product or package will break down into, or otherwise become part of, usable compost (e.g., soil conditioning material, mulch) in a safe and timely manner in an appropriate composting program or facility, or in a home compost pile or device."

Claims may be considered deceptive if municipal composting facilities are not available to a substantial majority of consumers or communities where the package is sold, the claim misleads consumers about the environmental benefit provided when the product is disposed of in a landfill, or consumers misunderstand the claim to mean that the package can be safely composted in their home compost pile or device when in fact it cannot.

Here are some examples.

A manufacturer indicates that its unbleached coffee filter is compostable. This claim is not deceptive provided the manufacturer can substantiate that the filter can be converted safely to usable compost in a timely manner in a home compost pile or device, as well as in a community composting program or facility.

A lawn and leaf bag is labeled "Compostable in California Municipal Yard Waste Composting Facilities." The bag does break down, but contains toxic ingredients that are released into the compost material. The claim is deceptive if the presence of these toxic ingredients prevents the compost from being usable.

Another lawn and leaf bag, marketed nationally, is labeled "compostable." Also printed on the bag is a disclosure that the bag is not designed for use in home compost piles. The bags are in fact composted in municipal yard waste composting programs in many communities around the country, but such programs are not available to a substantial majority of consumers where the bag is sold. This claim is considered deceptive since reasonable consumers living in areas not served by municipal yard waste programs may understand the reference to mean that composting facilities accepting the bags are available in their area. To avoid deception, the claim should be qualified to indicate the limited availability of such programs, for example, by stating "Appropriate facilities may not exist in your area." As a consumer, you should know if municipal composting facilities exist in your area before you purchase compostable products.

A manufacturer labels its paper plates to be suitable for home composting. If the manufacturer possesses substantiation for claiming that the paper plate can be converted safely to usable compost in a home compost pile or device, this claim is not deceptive even if no municipal composting facilities exist.

A manufacturer makes a claim that its package is compostable. Although municipal composting facilities exist where the product is sold, the package will not break down into usable compost in a home compost pile or device. To avoid deception, the manufacturer should disclose that the package is not suitable for home composting.

A disposable diaper bears the legend "This diaper can be composted where municipal solid waste composting facilities exist. There are currently [X number of] municipal solid waste composting facilities across the country." The claim is not deceptive, assuming that composting facilities are available as claimed and the manufacturer can substantiate that the diaper can be converted safely to usable compost in municipal solid waste composting facilities.

State Regulations

The state of Indiana defines compostable as a material that will decompose into a soillike material in less than one year under controlled biological circumstances.

Compostable Products

Almost any solid, biodegradable organic material can be composted, regardless of whether or not it is labeled as such. Grass clippings, leaves, wood chips, vegetable and fruit peels and cores, coffee grounds, tea bags, eggshells, soiled paper that can't otherwise be recycled (such as paper towels and napkins, and molded paper items such as paper plates and egg cartons) can all be composted. Avoid putting in meat, fish, or dairy products.

HOW TO MAKE YOUR OWN COMPOST PILE

Sometimes it seems there are more "recipes" for making compost than there are for chocolate chip cookies. At the very simplest, just placing the compostable material in a heap will work, but there are many methods and devices for speeding up the process.

The basic instructions for making a "real" compost pile are to alternate layers of fresh "green" material (such as food wastes and grass clippings) with dry "brown" material (such as dry leaves). Chopping the material up into small pieces helps it break down faster. Wet each layer thoroughly and keep the pile moist but not soggy. Sprinkle a little soil or manure between the layers to start the process. You can regularly add material such as kitchen scraps or plant prunings. Turn the pile every couple of weeks if you are in a hurry. When the materials have all broken down and are dark and crumbly, you have compost. Left alone it takes about a year to break down, but you can make it go faster by buying a tumbler or other bin, starter, thermometers, etc. Your local nursery probably has information, tools, and supplies for composting (see GARDEN).

C H A P T E R 1 2

Socially Responsible

To live in a sustainable world means sustaining not only the earth, but also its people; therefore, the human element of a product needs to be considered, too. A product can be made in a system of exploitation where a few top people make a lot of money at the expense of workers and the environment, or it can be made in a system where there is an attitude of cooperation, workplace democracy, responsibility, and shared wealth.

SOCIALLY RESPONSIBLE

For more than twenty years now the term *socially responsible* has been used to describe companies that support human values and contribute in some way to the resolution of social issues, rather than being exploitive or destructive. Even so, there still is no popularly agreed-upon or legally regulated definition. Rather it is a catchall term that can include not only people-oriented issues, but also animal rights and the environment.

The most popular guide to socially responsible products, the best-selling *Shopping for a Better World: The Quick and Easy Guide to Socially Responsible Supermarket Shopping*, gives detailed ratings in these major areas:

- Generous charitable giving—contributions of money (0.7 percent to 2 percent or more) or products to such social programs as food banks, job-

training programs for disadvantaged youths, community revitalization projects, and public education
- Equal opportunity—equal opportunity for jobs or advancement to people of ethnic minorities, women, or the disabled, or do business with businesses owned by these groups
- Ending animal testing—elimination or reduction of laboratory testing on animals
- Right to know—full disclosure of companies' social efforts
- Community outreach—programs promoting education, housing, or volunteerism and making investments in these programs
- Ending apartheid—boycott of business in South Africa to protest racial inequality
- Environment—substantial positive environmental programs, such as the use and encouragement of recycling, waste reduction, green products, etc. (the company also must have a record relatively free of major regulatory violations)
- Family benefits—programs that support families, such as flexibility in workplace policies and child care assistance
- Workplace—honors unions, no worker safety violations, good pension and medical plans.

THE CERES PRINCIPLES

The *Exxon Valdez* oil spill disaster prompted shareholder groups to organize the Coalition for Environmentally Responsible Economies (CERES), dedicated to using their economic clout to influence corporate behavior. CERES devised the Valdez Principles (now called the CERES Principles) which companies use as guidelines for becoming more environmentally responsible; you can also use these guidelines when choosing companies to buy products from or invest in. Keep in mind, though, that this is a pledge, not a description of a company's actual practices. If you are considering a company that has signed the CERES Principles, ask them how they are actually implementing them in their business. Signatories to the CERES Principles are required to complete an annual CERES Report which discloses information used to assess the real environmental impacts of their corporations.

Introduction: By adopting these principles, we publicly affirm our belief that corporations have a responsibility for the environment, and must conduct all aspects of their business as responsible stewards of the environment by operating in a manner that protects the Earth. We believe that corporations must not compromise the ability of future generations to sustain themselves.

We will update our practices constantly in light of advances in technology and new understandings in health and environmental science. In collaboration with CERES, we will promote a dynamic process to ensure that the Principles are interpreted in such a way that accommodates changing technologies and environmental realities. We intend to make consistent, measurable progress in implementing these Principles and apply them to all aspects of our operations throughout the world.

- *Protection of the biosphere.* We will reduce and make continual progress toward eliminating the release of any substance that may cause environmental damage to the air, water, or the earth or its inhabitants. We will safeguard all habitats affected by our operations and will protect open spaces and wilderness, while preserving biodiversity.
- *Sustainable use of natural resources.* We will make sustainable use of renewable natural resources, such as water, soils, and forests. We will conserve non-renewable natural resources through efficient use and careful planning.
- *Reduction and disposal of waste.* We will reduce and where possible eliminate waste through source reduction and recycling. All waste will be handled and disposed of through safe and responsible methods.
- *Energy conservation.* We will conserve energy and improve the energy efficiency of our internal operations and of the goods and services we sell. We will make every effort to use environmentally safe and sustainable resources.
- *Risk reduction.* We will strive to minimize environmental, health, and safety risks to employees and the communities in which we operate through safe technologies, facilities and operating procedures, and by being prepared for emergencies.
- *Safe products and services.* We will reduce and where possible eliminate the use, manufacture or sale of products and services that cause environmental damage or health or safety hazards. We will inform our customers of the environmental impacts of our products or services and try to correct unsafe use.
- *Environmental restoration.* We will promptly and responsibly correct conditions we have caused that endanger health, safety or the environment. To the extent feasible, we will redress injuries we have caused to persons or damage we have caused to the environment and will restore the environment.

- *Informing the public.* We will inform in a timely manner everyone who may be affected by conditions caused by our company that might endanger health, safety or the environment. We will regularly seek advice and counsel through dialogue with persons in communities near our facilities. We will not take any action against our employees for reporting dangerous incidents or conditions to management or to appropriate authorities.
- *Management commitment.* We will implement these Principles and sustain a process that ensures that the Board of Directors and Chief Executive Officer are fully informed about pertinent environmental issues and are fully responsible for environmental policy. In selecting our Board of Directors, we will consider demonstrated environmental commitment as a factor.
- *Audits and Reports.* We will conduct an annual self-evaluation of our progress in implementing these Principles. We will support the timely creation of generally accepted environmental audit procedures. We will annually complete the CERES Report, which will be made available to the public.

Signatories: Ally Capital Corporation, Atlantic Recycled Paper Company, Aurora Press, Aveda Corporation, B&B Publishing, The Beamery, Inc., Bellcomb Technologies, Inc., Ben & Jerry's, Bestmann Green Systems, Inc., The Body Shop, Calvert Social Investment Fund, Clivus Multrum, Inc., Community Capital Bank, Consumers United Insurance Company, Co-op America, Council on Economic Priorities, Coyote Found Candles, Inc., Crib Diaper Service, Cyclean, Inc., Domino's Pizza Distribution Corporation, Ecoprint, Eco-Logical Marketing, Easlen Institute, Falcon Partners Management, L.P., First Affirmative Financial Network, Franklin Research & Development Corporation, Global Environmental Technologies, Geo. W. King, Company, Greenworld Products Corporation, Harwood Products, Indian Foods Company, Intrigue Salon, LecTec Corporation, Metropolitan Swere District, MoneyMatters Corporation, Ltd., Pacific Partners International Investments, Performance Computer Forms, Phoenix Heat Treating, Inc., Progressive Asset Management, Ringer Corporation, Service Litho-Print, Seventh Generation, Smith & Hawken, Stonyfield Farm Yogurt, Sullivan & Worchester, The Summit Group, Sun Company, Inc., Tom's of Maine, VanCity Investment Services, Ltd., Walnut Acres, The WATER Foundation, Working Assets Funding Services.

In addition, *Shopping for a Better World* gives alerts for products made by companies that have military contracts, supply the nuclear power industry, have nuclear weapons-related contracts, are foreign-based, practice forest clear-cutting, make cigarettes, and make pesticides, and kudos to companies that give to 1% for Peace, are worker coops, have endorsed the environmental CERES Principles (see sidebar), endorsed the Workplace Principles describing the rights of employees with AIDS, and other similar issues.

All of these are valid and admirable reasons for choosing or rejecting a product; however, it's important to keep in mind that a socially responsible product (or investment offering) is not necessarily a sustainable product, nor does it particularly have anything to do with the environment—"socially responsible" is used in such a variety of ways, that it could mean anything from the workers being well-paid to the product not being tested on animals. If a product or company claims to be socially responsible, it is necessary to find out exactly how they are.

The September/October 1993 issue of *The Utne Reader* featured socially responsible business as its cover story, and asked leaders in the field to comment on what socially responsible business meant to them. Only two (out of twenty-one) questioned the merits of socially responsible business by its current standards and both called for sustainability as a new guideline for business. Interestingly enough, neither are actively involved in running a business at this time.

Paul Hawken, co-founder of the Smith & Hawken gardening catalog, stated:

> If every company on the planet were to adopt the environmental and social practices of the best socially responsible companies, the world would still be moving toward environmental degradation and collapse . . .
>
> There is a contradiction inherent in the premise of a socially responsible corporation: to wit, that a company can make the world better, can grow, and can increase profits by meeting social and environmental needs. It is a have-your-cake-and-eat-it fantasy that cannot come true if the primary cause of environmental degradation is over consumption. Although proponents of socially responsible business are making an outstanding effort at reforming the tired old ethics of commerce, they are unintentionally creating a new rationale for companies to produce, advertise, expand, grow, capitalize, and use up resources: the rationale that they are doing good.
>
> In order to approximate a sustainable society, we need . . . a system of commerce and production in which each and every act is inherently sustainable and restorative . . . Businesses will not be able to fulfill their social contract with the environment or society until the system in which they operate undergoes a fundamental change, a change that brings commerce and governance into alignment with the natural world from which we receive our life. There must be an integration of economic, biologic, and human systems in order to create a sustainable and interdependent method of commerce that supports and furthers our existence . . . Business [must be] willing to integrate itself into the natural world.

Doug Tompkins, co-founder of Esprit de Corp clothing company, said:

> Socially responsible business is an oxymoron . . . So far all I have heard is a lot of superficial discussion and very shallow attempts by business to set socially responsible examples . . . But I do not believe they have any real chance because the very nature of their business is such that they promote consumption that isn't vital . . . They are displacing community-based products.
>
> Business, in my mind, needs major rethinking but that will only come with a revolutionary shift in worldview when we begin conducting civilization on deep ecological terms and not economic terms. Until we forget high technology and take up the advanced technology of indigenous and land-based cultures who proved their model worked for thousands of years (ours failing in often less than a few generations), the goals we all seek—security, health, clean air, food, water, healthy forests and soils—will be a forlorn hope. To do this will demand serious reappraisals of our most basic assumptions . . . Only then, when business becomes a consequence of an organic and ecocentric worldview, will it be truly socially responsible.

I believe that the future definition of a socially responsible business will be that it is a sustainable business: It sustains the planet, the people who work for it, and its customers. It will provide high-quality products that are truly necessary for our survival, produced in a sustainable way, and provide jobs that can sustain its workers. In his book, *The Ecology of Commerce*, Paul Hawken suggests that sustainable businesses

- Replace nationally and internationally produced items with products created locally and regionally.
- Take responsibility for the effects they have on the natural world.
- Do not require exotic sources of capital in order to develop and grow.
- Engage in production processes that are human, worthy, dignified, and intrinsically satisfying.
- Create objects of durability and long-term utility whose ultimate use or disposition will not be harmful to future generations.
- Change consumers to customers through education.

I know of no socially responsible companies that completely meet these criteria today, but there are some consumer choices you can make that will support positive people-oriented issues, for an important part of building a sustainable world is to purchase products from companies where your purchases contribute to building your own local economy, and where workers are treated with respect and fairly compensated.

MADE IN THE U.S.A.

The most sustainable system of commerce is to have products made in close proximity to where the resources have been taken and where the product will be sold. At the very least, products should be made in the country where they are sold (and preferably the state or local region), so as to keep the local economies flourishing.

Many products on the market today are imported from other countries—they are made from foreign resources by foreign labor, and the profits end up in foreign pockets. Though these products may seem glamorous or prestigious, every time you buy a product not made in the United States, you are displacing a U.S. job and contributing to unemployment. One small example of how foreign trade hurts U.S. workers was in my local paper recently. Here in California we grow most of this nation's garlic. But China is now beginning to export garlic at much cheaper prices. Because labor costs are so low, even with shipping costs, the retail price of the garlic is lower than the California growers can compete with. There's no reason to export garlic from China to the United States when we can grow it here, especially when that import undermines our own domestic business and economy.

But don't assume that if a product is marketed by a U.S.-owned company that it is made in the United States—it is common practice today for major manufacturers of mass-market goods to make products in other countries because labor prices are cheaper. This also leads to unemployment here in the United States because of further displacement of jobs and because products made in the United States can't compete pricewise with those made by cheaper labor. While it may appear on the surface that these lower-priced goods are stimulating the economy, when too many of our jobs are displaced, we don't have the money to buy even basic necessities. To sustain our economy, we have to have our products made by our own workers and keep our money circulating around our own economy.

Another concern with U.S. companies employing foreign workers to manufacture our consumer goods is that foreign workers often suffer under conditions that are far below what we would consider acceptable in this country. Some of the problems include substandard wages, unsafe working conditions, long working hours, child labor, prison or forced labor, discrimination in hiring, and physical and mental abuse.

There has been a pattern over the last several decades of U.S. exploitation of workers in various foreign countries. The company goes into a country where labor costs are low, employs workers at extremely low wages, then as labor costs rise, the company pulls out and takes its business to another foreign country with cheap labor. While this quickly increases the local economy, it provides little stability for sustaining any country's economic base.

Third World labor fuels our Western consumer society, and provides the profits that are divided between the millions of shareholders who own pieces of U.S. companies. Instead of putting their money into paying U.S. workers, companies hire cheap foreign labor and spend the difference on advertising and marketing campaigns to get Americans to buy their products.

As only one example of this kind of Third World exploitation, here are excerpts from a recent article by Nina Baker of the Portland newspaper *Oregonian* that describes the working conditions in an Indonesian factory. This story happens to be about Nike shoes, but the business practices are not unique to them—they are common for U.S. companies.

> Over and over through the day, Tri Mugiyanti dabs paint on the soles of fresh-off-the-mold Nike sneakers. Around her on the production line, workers sit or stand elbow-to-elbow. Many are barefoot. The humid air reeks of paints and glues. The temperature hovers near 100 degrees. Breathing feels unnatural . . .
>
> Stay ten minutes in the factory where [Tri Mugiyanti] works and your head will pound, your eyes and lips will burn. Amid the glue and paint fumes, workers without protective clothing operate hot molds, presses and cutting machines. A rubber-room fire killed one worker last year . . .
>
> The factory, which cranks out 370,000 pairs of Nikes a month and aims to do more, must work into the night to keep up with demand. Compulsory overtime is illegal in Indonesia. But enforcement is lax, and Tri and her co-workers would lose their jobs if they refused the additional hours. Besides, it is hard to survive without the extra money.
>
> Tri . . . earns 15 cents an hour . . . It would take all of Tri's pay for seven weeks to buy one pair of the Nike shoes she helps make. Contrast her wages with those of basketball star Michael Jordan, Nike's $5 million a year pitchman. Nike will spend $180 million on advertising [in 1992]. [In 1991], Nike shareholders earned $329 million in profits on sales of $3.4 billion [a 39% profit margin].
>
> Even in Indonesia, where per capita income is $585 a year, the wages of shoe factory workers . . . are 35 percent below the government's standard for the minimum physical needs of a single adult. As in many developing countries, Indonesia's minimum wage, which is what Tri earns, is less than poverty level.
>
> [Nike] has nothing invested in manufacturing plants or equipment. Instead, the company—which makes no shoes in the United States—contracts with 35 independent shoe factories sprinkled like colonies throughout Indonesia, China, Thailand, South Korea, and Taiwan.
>
> Indonesian investors own only two of the six Nike-licensed factories around Jakarta . . . All six plants are managed by South Korean concerns, so Indonesian workers have little or no opportunity to advance. None of the machinery used to produce the shoes is made in Indonesia. And about 70 percent of the factories' raw materials comes from South Korea, where escalating labor costs have made it too expensive

> to make all but Nike's most technologically advanced and expensive shoes.
>
> Because one pilfered pair of Nikes yields the equivalent of a half-month's salary for workers who sell them on the street, security is tight. Guards stationed at the factory's doorways and gates frisk workers at the end of their shifts. Workers wear t-shirts that are color-coded by department so the Korean managers can better track their movements. Those caught stealing are sometimes marched through the factory with signs on their backs before they are fired.
>
> After a long day, Tri . . . stands outside her home and says she is very tired. But home is hardly a place to relax. She shares one room in a slum less than a mile from the factory with three other workers. Smoldering piles of garbage line the cracked dirt paths to her doorway. She sleeps on a bamboo mat.

In all fairness, the other side of this story is the U.S. companies that provide business for these foreign factories are helping them industrialize. In Japan, Korea, and Taiwan—which began manufacturing shoes and textiles under similar conditions twenty years ago—factories now produce cars and electronics. But still there is no reason I can think of why these companies couldn't take a bit off their profits or advertising budget to ensure safe conditions and fair pay for their workers. They don't do it because supply and demand don't force them to do it. But they could do it inspired by their own ethics and compassion (or consumer pressure).

Best to buy made in the United States, or even better, made in your own home state or local community, to keep our jobs here at home.

WORKERS' COOPS AND FAMILY-OWNED BUSINESSES

It is important for sustainability to support locally owned and family-owned businesses, and workers' coops. Such small businesses can make regionally appropriate products in adequate supply for a town or region that each give ample income to a few people or families, but that would never be profitable for a large corporation.

The corporation itself must come into question if our goal is sustainability. Corporations, by their very legal structure, are not sustainable. By federal law the management of publicly held companies have a fiduciary responsibility to act primarily in the economic interests of the shareholders, which means that corporations have to do what is best economically for the shareholders rather than what is best for the environment or society, or they can be sued by the shareholders (not much incentive here to put the environment or human welfare at the top of the list of priorities!). Their number-one concern is profits, profits, profits, and they will (and do) exploit workers and manipulate consumers for those profits.

The success of a corporation is measured solely by financial gain; therefore the most successful corporations are those that expand and expand, turning more and more natural resources into consumer goods. All other values are secondary. How can a corporation be socially responsible if its reason for being is to consume the earth, and to do so as fast and as completely as it can? This is not to say that it is impossible for a corporation to both be completely sustainable and profitable, but to do so, it would need to change its priorities, and that is very difficult with current legal dictates. Nor am I suggesting that we should boycott all corporations, but rather consider other options as we move toward sustainable business, and support noncorporate structures when we have the choice.

The alternative to large publicly held or privately held corporations are family-owned businesses and workers' coops. Unlike corporations, they can work at a local level to produce regionally appropriate products by hand or with small-scale technology, and the workers can have control over the business values and working conditions. Many employ home workers, who can stay with their children and still earn a living. These small businesses are vital to a sustainable economy and it is important to patronize them even if the prices are higher than mass-market discounters, to keep that money circulating in your local economy.

As a complement to family-owned and one-person businesses, workers' cooperatives provide a structure other than the corporation for people to work together and profit from their joint efforts. In parts of the United States and in the Third World, people who have nothing but their hands and their desire to work are forming small workers' coops and producing clothing, jewelry, baskets, food, and other items they can sell. Using inexpensive, renewable resources (that are often sustainably harvested), these workers create their own jobs and provide products filled with a spirit no mass-produced machine-made product can. By purchasing these products, you provide "seed money" (so to speak) for these communities to start developing their own local economies and, in the Third World, restore their own traditional sustainable life-styles.

There's no reason why this form of business should be limited to the poor; one of the biggest and most famous workers' coops is in the town of Mondragon in the Basque region of Spain. Since their beginnings in the 1940s, the worker coops in the region have grown to include 170 schools, houses, stores, clinics, foundries, robots, factories, and banks. Coops range in size from six to 2,000 cooperators, employing more than 21,000 workers. Their system is based on, to quote *We Build the Road As We Travel*, "a many-voiced, spirited, and democratic pursuit of . . . *equilibrio*. This means not just equilibrium or balance, but also implies harmony, poise, calmness, and composure. Equilibrio is a vital process that harmonizes and balances a diverse and growing community of interests: those of the individual and the co-op, the particular co-op and the co-op system, and the co-op system and the community and the environment." That's a sustainable statement of purpose if I ever read one!

It is small local businesses that will build local sustainable economies, by employing and providing sustenance to friends and neighbors and using local resources wisely. This kind of business is truly "socially responsible."

Part Two

Part Two contains guidelines for choosing the most sustainable products in more than one hundred product categories. This listing is not meant to be exhaustive either in product types or in listing available products or catalogs—it simply gives a general idea of what you'll find in the marketplace on gradient levels of sustainability. The guidelines are meant to open doors for you to explore for yourself, not to send you searching for a specific product.

There are so many products with environmental attributes available today that I cannot possibly list them all here, or as one person even know about all of them. In choosing product categories to address, I've given the most attention to those products we all use every day—food, clothing, cleaning products, and the like—and less detail to products such as pet care or gardening, for which other books have been written and which are not applicable to everyone. I also include only those product types for which I am aware of a more sustainable choice; I haven't included any product where the recommendation would simply be "avoid" because there are no alternatives, such as cigarettes or nail polish. My purpose here is to focus on the positive possibilities, not alert you to all the possible health and environmental dangers.

When giving brand-name products or mail-order catalogs as examples, I either list where to locate a hard-to-find item, or give a common example of a product that would be easy to find, no matter what part of the country you live in. I also list a few mail-order sources for every category; these are not the only catalogs, but they are well-established and offer a variety of products.

One of the difficulties in recommending sustainable products in a book such as this is that to evaluate a product for sustainability requires looking at both the product itself and its distribution—in addition, of course, to the needs of the user—so even if I do give a suggestion, you still need to compare that to what you find in your own local area. There are some products on the market today that come close to being fairly sustainable when all you consider is the process of making the product. But as soon as you transport that product across the country, it becomes less and less sustainable the further it goes from the factory because of the nonrenewable, polluting energy used in transportation, and less and less appropriate to the ecology of the place in which it is being used.

A perfect sustainable product would be

- made from local renewable resources grown, raised, and harvested in a sustainable way or from recycled material,
- crafted by local cooperatives in an environmentally sustainable way,
- sold to local citizens, and
- disposed of (or recycled) in the region.

In the practical reality of today's marketplace, however, what we find are products with widely varying degrees of sustainability. In evaluating products, you need to look at two things: product sustainability and the distance you are from its manufacture. Some local products are unsustainable in their manufacture and ingredients; some products made far away are much more sustainable in their composition and may therefore be the better choice. My general rule of thumb is to buy the most sustainable product regardless of how far it has been shipped (to help establish the market), but if I have the choice of two products of comparable sustainability, I'll choose the one closest to home. Checking the distance between you and the manufacturer is easy if the point of manufacture is noted on the label. Given that close proximity of manufacturer is a standard for choosing all products, the product listings that follow will only give the product specifications.

Choosing products for sustainability is not a yes or no proposition—sustainability occurs on a gradient. To take this into account, I've formatted the product listings as descriptions of products you can look for in your local stores or order by mail, with their relative sustainability indicated with a series of arrows (see Explanatory Notes).

FINDING SUSTAINABLE PRODUCTS

Though there have been a number of books published on environmental products, I still have to recommend my *Nontoxic, Natural & Earthwise.* Even though it's several years old, as I was writing this book I was pleasantly surprised to see how useful and up-to-date much of that information still is. Use it as a companion guide to this, particularly for more detailed product information, mail-order sources, and do-it-yourself formulas.

You can start right out, without even looking at the product listings, by ordering the major environmental "general merchandise" catalogs: ***Seventh Generation, Real Goods, Heart of Vermont, Earth Care,*** and ***Coop America.*** These carry a wide variety of products, rather than specializing in a particular product area. The main socially responsible catalog is ***Pueblo to People,*** which features colorful clothing, toys, jewelry, furniture, and other items made by cooperatives in nine Central and South American countries.

Next, check the Directory of Greenstores for an environmental store near you. If you don't find one on the list, ask your natural food store—more and more greenstores are opening all the time. Many natural food stores, too, now have sections for nonfood products, particularly cleaning products and cosmetics, and some even carry things like recycled paper and beeswax candles.

And browse through magazines that regularly write about or advertise environmental products. I've listed twenty-five such publications in Recommended Reading; check the magazine rack at your natural food store or greenstore for others.

EXPLANATORY NOTES

⬆ Up arrows indicate that the product described moves toward sustainability. It was my conscious intention to put the most sustainable choices first; however, sustainability gradients are not always linear because there are so many factors to consider.

⬇ Down arrows are only used if there are particular products that are exceptionally unsustainable, and I want to suggest they not be used. If there is no particular warning, I only list the more sustainable choices.

Mail-order catalogs are in boldface italic type and their addresses appear in the Directory of Mail-Order Catalogs at the end.

Brand names are in boldface type. Manufacturers' names can be found in the Directory of Product Manufacturers. Complete addresses are not given for manufacturers because generally you cannot purchase directly from them; however, if they do mail-order, the manufacturer's name is in boldface italics and complete contact information is given in the Directory of Mail-Order Catalogs. Many, however, will direct you to local outlets for their products and some have toll-free numbers (call 800-555-1212 for 800 number information). If a product is mentioned that is not carried in your local stores, it's best to ask the manager of your favorite store to order the product and stock it.

Air Conditioners

⬆ Natural Means. There are many ways to cool your home that are in harmony with the natural features of your land and your climate. In the vernacular architecture of the Middle East, for example, buildings had air scoops to catch the cool evening breezes. Sustainable cooling systems utilize the coolness of the northern exposure, shade trees, the temperature of the soil, flowing water, overhangs, and awnings to block sun from coming in windows, as well as open windows.

You can reduce your need for air-conditioning by minimizing the amount of heat that is created in your home. The three major sources of heat gain are heat that conducts through your walls and ceilings from outside air, waste heat that is given off inside your house by lights and appliances (do your baking and ironing in the evening when it's cooler), and solar heat gain from the sun shining through your windows. To minimize these heat sources, weatherize your home, use energy-efficient appliances and lights, and shade windows with vegetation or an interior window covering (contact the energy organizations listed in the Directory of Organizations for more tips on natural cooling).

♠ FAN. You can reduce your need for air-conditioning by using fans. A whole-house fan in your attic (or an upstairs window) can be used to pull cool air into your home and blow warm air out, saving up to two thirds of your cooling costs. Even if you have air-conditioning, it will pay to use the fan rather than air-conditioning when the outside temperature is below 78 degrees F. Using a solar-powered fan has the advantage of running on the sun's abundant, nonpolluting, and free energy exactly when you need it to—during the sunniest and hottest part of the day.

♠ ENERGY-EFFICIENT AIR CONDITIONER. If it is absolutely necessary for you to have an air conditioner, your options are to choose from one of three types: room air conditioners, central air conditioners, and electric heat pumps (see **HEATING SYSTEMS**). Room air conditioners are installed in a window or wall and sized to cool just one room, whereas central air conditioners are designed to cool the whole house.

Central air conditioners and heat pumps operating in the cooling mode are rated according to their seasonal energy efficiency ratio (SEER)—the seasonal cooling output in BTUs divided by the seasonal energy input in watt-hours for an average climate in the United States. The National Appliance Efficiency Standard for central air conditioners requires a minimum SEER of 10, though some models are available with SEER values of 13 to 16. The SEER value appears on the EnergyGuide sticker and the most efficient brands and models are listed in *Consumer Guide to Home Energy Savings.*

The efficiency of room air conditioners is measured by the energy efficiency rating (EER), which is the ration of the cooling output (in BTUs) divided by the power consumption (in watt-hours). It is different from the SEER in that it measures only the efficiency when the unit is running and does not factor in seasonal fluctuations. The National Appliance Efficiency Standards for room air conditioners vary depending on the design and cooling capacity of each unit; on average the minimum requirement is an EER of about 8.6, and an EER of 9 or more is considered efficient. The EER value appears on the EnergyGuide sticker and the most efficient brands and models are listed in *Consumer Guide to Home Energy Savings.*

↓ CFC COOLANTS. All home air conditioners are cooled by ozone-depleting CFCs—some newer models use the less harmful HCFCs. If you already have an air conditioner in good working order, it's probably better to keep it and use it as long as you can, rather than dump it. If you must buy a new air conditioner, choose one cooled with HCFCs.

Air Fresheners and Odor Removers

↑ FIND THE SOURCE OF THE ODOR AND REMOVE IT. Odors are often produced by molds and bacteria. Empty the garbage frequently, keep things clean, dispose of rotting vegetables.

↑ OPEN THE WINDOWS. Ventilation will dilute and remove many odors.

↑ MAKE YOUR OWN. Baking soda and the natural mineral zeolite (one brand is **NonScents**) will absorb odors, without adding fragrance to the air. You can also add a few drops of any essential oil to a pump spray bottle of water to make your own scented air freshener.

↑ ORGANICALLY GROWN OR SUSTAINABLY HARVESTED NATURAL/RENEWABLE INGREDIENTS. Look for herbal potpourris and sachets made with organically grown ingredients. One sachet sold in many gift shops is made by ***Clear Light***—crumbled, sun-dried sustainably harvested cedar boughs, packed in little cotton bags.

↑ NATURAL/RENEWABLE OR HYBRID-NATURAL INGREDIENTS. There are many herbal potpourris, essential oils, and natural air fresheners sold in natural food stores and gift shops. **Air Therapy,** a highly concentrated citrusy nonaerosol spray, is sold in most green catalogs and natural food stores.

↓ TOXIC/NONRENEWABLE INGREDIENTS. Most supermarket air fresheners contain many toxic ingredients and do nothing more than cover up the odor with another one, or interfere with your sense of smell. Do not use them.

See also **CLEANING PRODUCTS**.

Air Purification

↑ ELIMINATE THE SOURCE OF POLLUTANTS. The most sustainable choice would be to eliminate indoor air pollutants altogether, so there would be no need to collect toxic pollutants in filters and then dispose of them somewhere else in the ecosystem. Air-purification devices can reduce the amount of pollutants present in a closed indoor space, but don't assume that a machine can remove all pollutants or that the quality of the air is equivalent to fresh, clean outdoor air in a pristine place. Particleboard furniture and cabinets, cleaning products, pesticides, plastics and synthetic fibers used in furnishings and other construction, carpeting, drapes, scented items, gas appliances and heaters, and many other common items made from petrochemicals all contribute to an increase in indoor air pollution.

🠅 VENTILATE. Just opening a window will allow many pollutants to escape. Opening two windows on opposite sides of a room or house will create a cross-draft that will clear the air faster and more completely. If you need more ventilation but don't want to lose heat, use an AIR-TO-AIR HEAT EXCHANGER.

🠅 PLANTS. Tests done by NASA have shown that common houseplants remove pollutants as they go through their natural process of photosynthesis—while plants draw in carbon monoxide, they also pick up airborne pollutants through small openings called stomates in the leaves. They are very effective at removing gases such as formaldehyde, carbon monoxide, carbon dioxide, benzene, cigarette smoke, and ozone, which are harmful for us to breathe, but a gourmet meal for a plant (such is the mutualism of nature). Aloe vera (which is a good plant to have around to treat burns and skin irritations), bamboo palm, common chrysanthemums, dracaena palms, philodendrons, golden pothos, spider plants, and scheffleras make the best air filters. It takes one or two good-sized plants to purify the air in a ten by ten-foot room, depending on the level of pollutants present.

🠅 CARBON AND/OR HEPA AIR FILTER FOR REMOVING VOLATILE GASES AND/OR PARTICLES. Carbon alone will remove gases (misty vapors of volatile chemicals such as formaldehyde, plastics, paints, solvents, pesticides, and perfumes), HEPA alone will remove particles (bits of pollen, dust, mold, and animal dander); together they give the highest efficiency broad-spectrum removal.

Activated carbon used in filters works by adsorption, a process by which pollutant gases are attracted by and stick to the carbon. There are several types of activated carbon; coconut shell carbon is generally considered to be the highest quality. Filters can also contain other filter media that have special qualities. Some coconut-shell carbon is impregnated with nonmetal salts to increase efficiency in removing formaldehyde—up to 90 percent. Carbon can also be combined with Purafil, a nontoxic odoroxidant made of activated alumina impregnated with potassium permanganate. Purafil works by both absorbing and adsorbing gases and then destroying them by oxidation. This combination is more effective that carbon alone, but not as effective as impregnated carbon.

High-Efficiency Particulate Arrestance (HEPA) filters work by trapping particles mechanically. They are rated at 99.99 percent efficiency for particles 0.3

microns in size (dust, pollen, and plant and mold spores). Developed by the Atomic Energy Commission during World War II to remove radioactive dust from industrial exhausts, they are paperlike filters made of randomly positioned fibers that create narrow passages with many twists and turns. As the air passes through, particles are trapped, clogging holes and making the grid smaller, which enables the filter to be even more efficient with ongoing use.

Activated carbon and other filter materials must be changed regularly. How often you'll need to change them depends on how many hours a day the filter is used and how polluted the air is, so it is impossible to predict how long filter media will last. Manufacturers estimate that activated carbon will last about two thousand hours, or twelve hours per day for six months; under normal use the carbon should last six to nine months. Prefilters are generally changed more frequently, and HEPA filters last for several years. If you are using a filter under unusual conditions, you'll need to adjust the expected life accordingly.

Air filters can be purchased as portable models, or you can have them built into your central heating–air-conditioning system. An important point to remember is that the cleaning capabilities of any portable unit are generally limited to the air in the room in which it is being used. And as with air-conditioning, air cleaners are most effective if no outside air is entering from adjacent rooms, open windows, or central air systems.

Each air purification unit is designed to effectively remove pollutants from the air contained in a certain measured space. The amount of filter media the unit contains, the rate at which the air flows through the media, the size of the motor, and all other aspects of the unit's design are geared to the designated room size. You may need several filters to continuously clean the air throughout the house, or a unit portable enough to be easily moved from room to room. The small air cleaners found in most department, drug, and discount stores are inexpensive and convenient, but do not contain sufficient carbon to effectively clean the amount of air that passes through.

Truly effective portable air filters are rarely sold in stores. Some manufacturers of quality portable air filters that I have known of and recommended for many years include ***Aireox Research Corporation, E. L. Foust Company,*** and ***AllerMed Corporation.*** Units can be purchased directly from the manufacturers, or from independent dealers such as ***Nigra Enterprises*** or ***Baubiologie Hardware.*** For a built-in whole-house filter, check the Yellow Pages for a Heating, Ventilation, and Air Conditioning (HVAC) dealer.

🡻 NEGATIVE ION GENERATORS. As air cleaners, negative ion generators are quite limited—they will precipitate only certain small particles. While practically useless for dust or pollen, they are very effective at removing the particles found in cigarette smoke and smog, cleaning the air so that it becomes clear and odorless. But they cannot remove the invisible, odorless toxic gases that are also present in cigarette smoke and smog. Negative ion generators and ionizers should be purchased for their health benefits or for use with activated carbon filters for removal of cigarette smoke, but not as broad-spectrum air cleaners. If you do purchase an ionizer, choose one with a built-in particle collector, so you won't end up with black particles stuck to your walls.

⬇ ELECTROSTATIC FILTERS. Electrostatic filters attract particles by electricity—either by way of an electronic air cleaner (electrostatic precipitator) or with electronically charged plastic fibers (electret). Generally neither type is recommended. Electrostatic precipitators are rarely more than 80 percent efficient and can quickly drop to 20 percent efficiency. They also produce ozone and positive ions, and must be cleaned often with volatile petrochemical solvents. Electret, on the other hand, is extremely efficient for removing particles, but is a petrochemical product and gives off a strong odor.

Air-to-Air Heat Exchangers

Air-to-air heat exchangers can effectively ventilate a closed house with only 20 to 30 percent heat loss. This machine blows stale indoor air out of the house, which passes into close contact with fresh outdoor air that is simultaneously being pulled in. This contact, occurring in many small, thin-walled tubes or channels, allows much of the indoor air's heat to be transferred to the incoming cold air so that warmth is retained.

Air is blown through a heat exchanger core that is made of metal, plastic, or treated paper. Tests at the Lawrence Berkeley Laboratory in California have found the metal core to work best at transferring heat (and it also would emit the fewest pollutants).

Heat exchangers are available in several sizes, from a window unit to whole-house systems that may require the installation of ducting. Depending on the amount of pollution being produced, it could take up to five or six small units to adequately ventilate an average house. You should be able to get more information on heat exchangers from a local HVAC contractor. Look in the Yellow Pages of your telephone book under Heating, Ventilation, and Air Conditioning.

All-Purpose Cleaners

⬆ MAKE YOUR OWN. Mix one teaspoon liquid SOAP into one quart warm or hot water. A squeeze of lemon or a splash of vinegar will help cut grease.

⬆ ORGANICALLY GROWN NATURAL/RENEWABLE INGREDIENTS. Not yet available to my knowledge, but watch for them.

⬆ NATURAL/RENEWABLE OR HYBRID-NATURAL INGREDIENTS. Many available in natural food stores and green catalogs.

⬆ NONTOXIC/NONRENEWABLE INGREDIENTS. Many available in natural food stores, green catalogs, and hardware stores.

⬇ TOXIC/NONRENEWABLE INGREDIENTS. Don't use products that contain ammonia, artificial dyes, detergents, or artificial fragrances.

See also CLEANING PRODUCTS.

Antiperspirants and Deodorants

⬆ MAKE YOUR OWN. The best deodorant I've found is just plain baking soda. This works better than anything else, even for those (I'm told) who have a problem with perspiration odor. If you want, you can mix baking soda with cornstarch (to make it softer) or fragrant dried herbs (for a subtle scent). After a bath or shower, just sprinkle baking soda onto your fingertips and apply to your underarms.

⬆ ORGANICALLY GROWN INGREDIENTS. A few natural deodorants—such as **Weleda Natural Sage Deodorant** (at your natural food store or order from ***Meadowbrook Herb Garden***)—are made from organically grown, biodynamically grown, or wildcrafted herbs.

⬆ MINERAL CRYSTALS. In the last few years, natural mineral deodorants that look like large crystals have become popular. They are sold in natural food stores and in many mail-order catalogs.

⬆ NATURAL/RENEWABLE AND HYBRID-NATURAL INGREDIENTS. Every natural food store should have a good selection of these.

⬇ HARMFUL INGREDIENTS. Aluminum chlorohydrate, the active ingredient in most popular brands of antiperspirant, can cause skin rashes.

See also **BEAUTY AND HYGIENE.**

Art Supplies

There are two types of art materials—the usually less toxic consumer products designed for general use, and the frequently more toxic professional or industrial

products intended for use by trained professionals in controlled environments. Both are widely available in art and office supply stores.

The labeling of art materials is supervised by the Consumer Product Safety Commission (CPSC), which administers three laws: the Labeling of Hazardous Art Materials Act of 1988 (an amendment to the Federal Hazardous Substances Act), the Poison Prevention Packaging Act of 1970, and the Consumer Product Safety Act of 1972. Under the Labeling of Hazardous Art Materials Act, art and craft manufacturers are required to report products that have the potential to cause chronic illness and to place labels on those that do, and in October 1992 the CPSC issued guidelines for determining those hazards. The CPSC may also require precautionary labeling of hazardous substances or ban substances for which labeling is determined to be insufficient to protect public health. Despite these laws many art materials have insufficient labeling, which is a result not of inadequate regulations but from lack of enforcement of these laws. California, New York, Massachusetts, Illinois, Tennessee, and Oregon all have more detailed state laws that attempt to improve art-product labels.

♦ Made from Natural/Renewable Ingredients. Natural beeswax crayons (and other charming children's toys and crafts) can be ordered through the ***Hearthsong*** catalog.

♦ Made from Nontoxic/Nonrenewable Ingredients. More than thirty companies have joined the Arts and Crafts Materials Institute, an industry group that has developed voluntary standards for the safety and quality of both consumer and professional/industrial art supplies. In evaluating the safety of a product, the institute's toxicologist considers both the concentration and the potential acute and chronic health effects of the ingredients, as well as possible uses and misuses of the products.

Manufacturers of safe products are allowed to pay the institute a fee to display certain certification seals such as the CP Nontoxic (Certified Product) seal, the AP Nontoxic (Approved Product) seal, and the Health Label. These seals represent that the product has been "certified in a program of toxicological evaluation by a medical expert to contain no materials in sufficient quantities to be toxic or injurious to humans or to cause acute or chronic health problems."

AP and CP Nontoxic labels are used on children's art supplies. In addition to the toxicological certification, CP products also meet specific requirements of material, workmanship, working qualities, and color standards. The Health Label is used on adult art materials to assure that they are properly labeled with health and use information, and includes a line that states "Nontoxic" or "Warning: Contains [name of the hazardous substance]." One drawback to this system is that the toxicologist's assessment of the safety of a product is based on a literature review, and not actual testing of the product itself.

Products with AP Nontoxic, CP Nontoxic, or the Health Label are available wherever art materials are sold. For a list of certified products, contact the Art and Craft Materials Institute.

Don't make the mistake of choosing a product without a warning label over a product with a warning label, assuming that no warning label implies that no warning is necessary. Whenever possible choose products that list ingredients or give some other clear indication of their safety.

🡇 MADE FROM TOXIC/NONRENEWABLE INGREDIENTS. To avoid the more stringent regulations for consumer product labeling, some product manufacturers label their more toxic products "for professional use only" or "for industrial use only." It is wise to stay away from these. Also avoid products that have complex instructions for providing adequate ventilation—another good indicator of toxicity.

Automobiles

🡅 DRIVE LESS. Automobiles are powered by nonrenewable fuels and create air pollution. Whenever possible, walk, bicycle, ride a horse, row a boat, take public transportation, share a ride, or carpool.

🡅 ELECTRIC AUTOMOBILES. Much research is being done on electric cars as the answer to our air pollution problems. In California the state has ordered that 2 percent of all vehicles sold by major automakers produce zero emissions beginning in 1998, with the requirement rising to 10 percent of vehicles in 2003. Electricity is the only power source that meets the zero-emission requirement. New York and Massachusetts have adopted identical standards, and they are also being considered by about ten other states.

An added bonus over zero emissions is that electric vehicles are more energy-efficient than gas-powered vehicles. A study by the Electric Power Research Institute in Palo Alto, California, compared two electric vans with two gas-powered vans. The gas-powered vans used 14,400 BTU/mile and 9,000 BTU/mile, where the electric vans used 10,800 BTU/mile and 5,400 BTU/mile—a savings of 25 and 40 percent, respectively. This includes the losses from mining, energy conversion, and the transmission of electrical energy for the electric vehicles. Electric engines make much more efficient use of energy than internal combustion engines, so when renewable sources of energy are used, driving an electric car could have very little environmental impact.

The biggest limitation to electric cars is their limited range. The best right now is about sixty miles—for many people electric cars make the perfect second or commuter car. Soon recharging stations will be as ubiquitous as gas stations.

Nearly all the world's manufacturers have launched electric car research programs. The most publicized has been the two-seat **Impact** made by General Motors. Expected to be available in the mid-1990s, it will have a top speed of 75 miles per hour and be able to go 120 miles without recharging. But there is no need to wait; you can convert a car to electric yourself with parts and plans from ***Electro Automotive.***

♠ FUEL-EFFICIENT AUTOMOBILES. Both the Environmental Protection Agency and Consumers Union (in *Consumer Reports* magazine) report yearly on the fuel efficiency of new cars. In addition, all new cars have the EPA gas mileage ratings posted right in the window. If you buy a used car, go to the library and look up the mileage rating in publications from past years.

Fuel economy figures are determined in tests conducted by the EPA to certify that vehicles meet federal emissions and fuel economy standards. Professional test drivers "drive" test vehicles under carefully controlled laboratory conditions on a treadmill-type device to ensure that all vehicles are tested under identical conditions. Two different tests are used, one to determine city driving mileage and one for highway mileage. Of course, actual mileage on the road will differ from laboratory tests, but the numbers can be used to compare the energy-efficiency of different models. The EPA mileage ratings for 1993 list **Geo Metro, Honda Civic,** and **Suzuki Swift** models in its top ten.

♠ RECYCLED/RECYCLABLE, ENERGY-EFFICIENT, ALUMINUM AUTOMOBILES. Currently 5 to 6 percent of autos are aluminum, more than 60 percent of the aluminum used in cars is made from recycled metal, and more than 85 percent of the aluminum used in cars is recovered and recycled. Use of aluminum by the auto industry has more than doubled from the average 77 pounds per car in 1971 to 160 pounds in 1992. Because aluminum is lightweight, the more aluminum used, the lighter the car is and the better the gas mileage.

The **Honda Acura NSX,** introduced in September 1990, is the most aluminum-intensive car on the road today and was the first to utilize aluminum in a broad and systematic way. The estimated total weight savings from using aluminum is about 450 pounds, which allows it to get a higher fuel economy than any other car in its class with similar performance capabilities.

♦ CFC-FREE AUTO AIR CONDITIONERS. Ford, General Motors, and Chrysler replaced CFCs with an HCFC coolant in some models beginning with 1993 models. By the end of 1993, all Nissan and Infiniti cars sold in North America also came with HCFC coolant.

If your car has an air conditioner, try not to use it. If you do, replace the air-conditioner hose in your automobile every three years. For repairs find an auto mechanic who has coolant-recycling equipment.

You can reduce or eliminate the need for air-conditioning by choosing a car with light exterior and interior colors and side vent windows, and you can add window glazings that slow solar adsorption, sunroofs, and solar ventilation systems.

B

Babies and Children

Infants' and children's products are in many ways smaller versions of their adult counterparts. The sustainability of these products is the same, regardless of the age of the user.

You'll find some natural baby care products and organically grown baby food at your natural food store (or you can make your own); however, most other items you won't find in a local baby store unless you have one that is exceptionally environmentally aware. My best suggestion is to order the ***After the Stork, Baby Bunz & Company, Biobottoms, Hanna Anderson, Motherwear,*** and ***Natural Baby Company*** mail-order catalogs and subscribe to *Mothering* magazine—these sources will lead you to everything you need. ***Seventh Generation*** catalog also has a lot of baby items.

Barbecue

♦ NATURAL WOOD BRIQUETS. Made from seasoned clean-burning wood (the first briquets were made from charred scrap wood left over from making the frames of Model T Fords). **Barbecubes Natural Fruitwood Briquets** (certified to be reclaimed by Scientific Certification Systems) are made from the prunings of fruit trees from the Central Valley of California, which are routinely pruned to encourage abundant fruiting the following season. Standard practice is to burn the prunings green in the field, which wastes this valuable source of energy and causes air pollution. Instead of field burning, this wood is collected, dried (using waste heat), chipped, and reformed to make Barbecubes. Available at natural food stores, supermarkets, and mass market retailers.

🡅 CHIMNEY-TYPE STARTER. This chunky cylinder with a handle is not only infinitely reusable, it also works better than any other kind of starter. According to the South Coast Air Quality Management District in Los Angeles, using a chimney starter to ignite briquets produces fewer air pollutants than any other method tested. Available almost everywhere barbecue supplies are sold.

🡇 CHARCOAL LIGHTER FLUID. Lighter fluid is made from toxic petroleum distillates, which produce VOCs that create smog. Because of this, lighter fluids and presoaked briquets have been put under strict regulation in particular areas, such as Los Angeles. According to law retailers are prohibited from selling lighter fluid if it emits more than 0.02 pounds of VOC pollutants per use. Already this prohibition has been expanded to apply to the entire state of California, and it is likely that other areas will follow. Most manufacturers of lighter fluid have reformulated their products to meet the standards. While this is an improvement, lighter fluids are still made from toxic substances derived from nonrenewable resources.

🡇 CHARCOAL BRIQUETS. Made by combining coal (a nonrenewable resource), limestone, borax, sodium nitrate, and sawdust with charred wood. Burning gives off primarily carbon monoxide, nitrogen oxide, and sulfur dioxide (which causes acid rain), as well as particulates and some organic compounds.

Basin, Tub, and Tile Cleaners

🡅 MAKE YOUR OWN. Baking soda works great, or use a nonchlorine SCOURING POWDER.

🡅 ORGANICALLY GROWN NATURAL/RENEWABLE INGREDIENTS. Not yet available to my knowledge, but watch for them.

🡅 NATURAL/RENEWABLE OR HYBRID-NATURAL INGREDIENTS. Not yet available to my knowledge, but watch for them in natural food stores, greenstores, and green catalogs.

♠ NONTOXIC/NONRENEWABLE INGREDIENTS. Some available in natural food stores, green catalogs, and hardware stores.

♣ TOXIC/NONRENEWABLE INGREDIENTS. Don't use products that contain aerosol propellants, ammonia, detergents, and artificial fragrances.

Descale-it Bathroom Cleaner, Descale-it Lime-Eater Bath & Kitchen Cleaner, Eliminate Shower Tub & Tile Cleaner, and **J. R.'s Tub & Shower Cleaner** are certified biodegradable by Scientific Certification Systems.

See also CLEANING PRODUCTS.

Bath Linens

♠ NATURAL FIBERS. One hundred percent cotton bath towels and bath rugs are sold in every department store in a wide variety of styles and colors.

♠ ORGANICALLY GROWN COTTON. Not yet available to my knowledge, but it's only a matter of time. As more organically grown cotton becomes available, it will be used to make bath linens, too.

♠ "GREEN COTTON." Unbleached, undyed cotton bath linens are carried wherever cotton towels are sold. Fieldcrest's **New World** line, sold in many department stores, has beautiful bath linens made with brown **FoxFibre** cotton on one side and unbleached cotton on the other.

See also TEXTILES.

Bath Products

♠ MAKE YOUR OWN. You can make any bath more luxurious by adding things you already have in your kitchen: milk and honey, lemon or grapefruit slices, chamomile or mint tea, rose petals or other fragrant flowers.

♠ ORGANICALLY GROWN INGREDIENTS. **Alexandra Avery** has biodynamically grown bath herbs (at your natural food store or by mail). Also **Dr. Hauschka** and **Weleda** bath oils (at your natural food store or by mail from ***Meadowbrook Herb Garden***) contain biodynamically grown herbs.

⬆ NATURAL/RENEWABLE AND HYBRID-NATURAL INGREDIENTS. Your local natural food store will carry plenty of these.

See also BEAUTY AND HYGIENE.

Batteries

⬆ RECHARGEABLE. Nickel cadmium ("ni-cad") batteries can be recharged nearly a thousand times—using a solar-powered battery recharger (available from ***Real Goods***), you can use entirely renewable energy. Ignore the new rechargeable alkaline batteries; though their initial cost is less, they can only be recharged twenty-five times. Better to make the initial investment and save money and resources in the long run.

⬆ LOW-MERCURY OR MERCURY-FREE DISPOSABLE BATTERIES. Virtually all the major battery manufacturers have reformulated their batteries to reduce or eliminate mercury. Look for the words "low-mercury" or "mercury-free" on package labels wherever batteries are sold.

Beans

⬆ ORGANICALLY GROWN. Easy to find in natural food stores and catalogs. Buy in bulk. There are also brands of canned cooked beans that use organically grown beans.

⬆ NATURAL. Available at your supermarket in plastic bags. Better to buy them at the natural food store in bulk to save on packaging and buy organic. Some national brands of cooked beans are additive-free—read the labels.

See also FOOD.

Beauty and Hygiene

Since 1977 the FDA has required a complete listing of ingredients on all domestic cosmetics, itemized in decreasing order and using standardized language. Some items commonly considered by consumers to be cosmetics, however, do not need to have their ingredients listed at all, because they are actually considered by the FDA to be over-the-counter drugs. According to the FDA definition a "cosmetic" is anything that can be "rubbed, poured, sprinkled or sprayed on, introduced into, or otherwise applied to the human body . . . for cleansing, beautifying, promoting attractiveness, or altering the appearance without affecting the body's structure or functions." If a product claims to affect the body's structure or function (such as fighting tooth decay), it is considered an over-the-counter drug. Hygiene items not covered by the cosmetics labeling requirements include deodorant soaps, fluoridated toothpastes, antiperspirants, sunscreens, and antidandruff shampoos.

And on those cosmetic items that do list their ingredients, everything is not necessarily revealed on the label. "Trade secrets" such as fragrance or flavor formulas are not divulged, and hide behind their standardized terms. Nor do the labels tell you whether or not the ingredients in "natural" beauty products are actually derived from natural sources.

🡅 MAKE YOUR OWN. Basic hygiene doesn't require anything fancy. There are lots of recipes for "kitchen cosmetics" in books available at your natural food store and local library.

🡅 ORGANICALLY GROWN INGREDIENTS. A few beauty and hygiene products are available—mostly by mail—made from organically grown ingredients.

🡅 NATURAL/RENEWABLE AND HYBRID-NATURAL INGREDIENTS. Natural cosmetics, made primarily from plant, animal, and mineral ingredients, are big business. In the last fifteen years, natural cosmetics have gone from being a few poorly formulated items in health food stores to having selection and quality that can be sold not only in natural food stores, but in department stores and specialty boutiques as well.

Because labeling laws require ingredients to be listed, if you know your ingredients it's a relatively simple task to identify natural cosmetics. Unfortunately, more than three thousand different ingredients are used in the manufacture of cosmetics, derived from petrochemical, animal, vegetable, and mineral sources, and there is no easy rule of thumb to give you to help identify the natural ones. The best advice I can give is to start reading labels and looking up the ingredients in books such as *A Consumer Dictionary of Cosmetic Ingredients*. Remember, though, that some of these ingredients may be hybrid-natural containing petrochemicals, even though they may say they have a natural source. Two nonrenewable petrochemical derivatives that are practically inescapable—even in natural cosmetics—are methylparaben and propylparaben. Laboratory tests have proved these common preservatives to be safe; however, I wouldn't call them natural.

Some clever manufacturers create "natural" formulas by adding natural-sounding ingredients such as honey or herbs or aloe vera, instead of actually making a more natural formula by removing unnecessary artificial colors, fragrances, and preservatives. And frequently I have found ingredients made from petrochemicals—particularly artificial colors—in cosmetics marketed "cruelty-free." Just because a product wasn't tested on animals doesn't mean it's natural, though they are marketed side-by-side with natural cosmetic items.

Here's an example of the ingredient list on a natural cosmetic product, **Dr. Bronner's Peppermint Oil Soap,** a mainstay that has been sold for years in every natural food store: Coconut Oil, Olive Oil, Peppermint Oil, Potassium Hydroxide, Water. All these ingredients are easily recognizable except maybe potassium hydroxide, which is a natural mineral salt.

Here's another one, **Tom's of Maine Toothpaste,** sold now in many drugstores and supermarkets as well as natural food stores: Calcium Carbonate, Sodium Bicarbonate, Glycerin, Sodium Lauryl Sulfate, Carrageenan, Peppermint Oil, Myrrh, Propolis. Calcium carbonate and sodium bicarbonate are both minerals. Glycerin is a by-product of soap making; while it can also be made from petrochemicals, generally in natural cosmetics it is made from an animal or vegetable source. Sodium lauryl sulfate is a common ingredient in many natural toothpastes, shampoos, and bath products. It is made by combining laurel alcohol (from coconut oil) with sodium sulfate (a mineral), followed by neutralization with sodium carbonate (a mineral). Carrageenan is extracted from seaweed, peppermint oil is obviously from peppermint, myrrh is a fragrant herb, and propolis is the tree resin collected by bees to line their hives. Once you start to become familiar with these natural ingredients, you'll see them used over and over in natural cosmetics.

Start by looking for natural cosmetics in your local natural food store. Most have a good selection of everything from soap, shampoo, and toothpaste, to complete makeup collections. These products can be reasonably relied upon to be made primarily from plant, animal, and mineral ingredients, though few are completely petrochemical free and the plant ingredients are grown with pesticides. Many department stores are also now selling natural cosmetics. Read the labels very carefully as I have found a number of petrochemical ingredients in them, particularly artificial colors and mineral oil. If in doubt about an ingredient, look it up. A little extra time is worth the effort, and once you've found brands that you like, you will also have learned to recognize ingredients.

🡇 MADE FROM POTENTIALLY HARMFUL INGREDIENTS OR THOSE DERIVED FROM NONRENEWABLE PETROLEUM. Space does not permit a complete listing of ingredients to avoid, but here are the major ones that are made from petrochemicals or are otherwise harmful to health or the environment: aerosol propellants, alcohol, ammonia, artificial colors and flavors, BHA/BHT, EDTA, ethanol, fluoride, formaldehyde, "fragrance," glycerol, glyceryl, hexachlorophene, isopropyl alcohol, methyl ethyl ketone, mineral oil, paraffin, phenol, anything that begins with PEG- or PPG-, PVP, quaternium 15, saccharin, and talc.

• • •

There are a few general personal care catalogs that I especially want to mention.

Simmons Handcrafts carries some of the purest natural bodycare products, including vegetarian olive oil soap they make themselves (they have recently added some other household items as well). Their catalog is printed on recycled paper and their entire home-based business runs on a renewable energy system.

Kettle Care offers bodycare products handmade with organically grown herbs from original recipes. This is also a home-based business—herbs come straight from the garden into the products. They pay a rebate when customers return empty containers.

Dry Creek Herb Farm, another family-owned home-based business, also grows its own herbs organically and makes a variety of personal care products.

See also ANTIPERSPIRANTS AND DEODORANTS; BATH PRODUCTS; COSMETICS; FRAGRANCES; HAIRSPRAY; MOUTHWASH; SHAMPOO AND CONDITIONER; SHAVING CREAM; SKIN CARE; SKIN LOTIONS, CREAMS, AND MOISTURIZERS; SOAP; SUN PROTECTION; and TOOTHPASTE AND TOOTHBRUSHES.

Bed Linens

⬆ NATURAL FIBERS. There are many cotton sheets available; however, it is important to make sure that they do not have formaldehyde resin finishes (see below). Cotton percale sheets often have a "no-iron" finish, while cotton damask and cotton flannel generally do not (cotton flannels often say "no-iron" on the label, but this is a feature of the fabric, not a finish). Natural bed linens are also available made from linen and linen-cotton blends. The least expensive cotton flannel sheets I've found can be ordered from ***Clothcrafters***; on the other end of the monetary spectrum, ***Chambers*** catalog carries the finest natural fiber bed linens from around the world. ***Garnet Hill*** is one of the original natural fiber bedding catalogs, with a good selection of nice bed linens. Most good department stores now carry natural fiber bed linens at affordable prices.

⬆ ORGANICALLY GROWN COTTON. Though expensive, bed linens made from organically grown cotton are available. ***Heart of Vermont*** carries flannel bed linens made from 100 percent organic cotton—brown or green **FoxFibre Colorganic** stripes on a creamy background.

⬆ "GREEN COTTON." The major manufacturers of textiles in this country are now coming out with affordable lines of unbleached, undyed, untreated cotton percale bed linens. Look for them in department stores and specialty shops. I've seen J. P. Stevens' **Simply Cotton** bed linens in department stores around the country, packaged as sets in recycled paperboard boxes. **New World** bed linens by Fieldcrest combine naturally colored brown **FoxFibre** and regular cream-colored cotton in a warm brown striated fabric.

↓ FABRIC FINISHES. Formaldehyde-based permanent-press or no-iron finishes are particularly a problem on bed linens. One of the symptoms commonly associated with formaldehyde exposure is insomnia, so if you can't sleep, you might try changing to "green cotton" sheets.

See also TEXTILES.

Beds

The most sustainable bed is one made from natural, renewable materials, the complete opposite of what most beds are made of today. There are two routes to take: buying a cotton mattress and box spring, or custom constructing a natural bed from layers of natural materials to suit your needs. I recommend the latter—not only can you have whatever degree of softness and support you like, but you can take the bed apart to air and sun the different pieces (natural fibers can collect moisture, and need periodic freshening). Here are the component parts that can be used to construct a natural bed.

↑ WOODEN SLAT BEDS. These provide a base for natural mattresses and allow air to circulate through the bedding. They are available in all futon shops and catalogs.

↑ FUTONS. These Japanese folding mattresses filled with cotton or wool batting have become quite popular in recent years. Most major cities now have futon shops, and often they are willing to make futons for you using your preferred materials. ***Jantz Design*** fills futons with organically grown cotton and wool; ***Heart of Vermont*** makes cotton and wool futons that are organic inside and out.

Futons are available in several thicknesses. Experience has shown that it's better to layer several thinner futons than have one thick one—in addition to being able to air them out, there is less shifting and lumping, and they are easy to fluff up (cotton futons can become quite hard after you've slept on them for a while).

The Federal Bureau of Home Furnishings requires that all futons be treated with a boric-acid flame retardant; however, most futon makers will be happy to make you an untreated futon upon presentation of a doctor's prescription.

↑ FEATHER BEDS. These are large pillows the size of a mattress, filled with feathers and covered with all-cotton downproof ticking. Not thick enough to be used alone as a mattress, feather beds are usually placed on top of a mattress or futon for added softness and warmth.

🠅 COTTON MATTRESS AND BOX SPRING. If you want a traditional mattress and box spring, it's better to have a cotton one than one made completely of nonrenewable, nonbiodegradable synthetic fabrics. They're not generally sold in stores, though you might be able to get a local mattress-maker to make one for you. If not, you can order them by mail from ***Janice Corporation.*** An all-organic cotton and wool mattress and boxspring is made by ***Heart of Vermont***.

See also TEXTILES.

Beer

🠅 ORGANICALLY GROWN. There are a few brands of beer made from organically grown ingredients—look for them at your natural food store.

🠅 NATURAL. There are many natural beers. All German beers are protected by a law called the *Reinheitsgebot* ("law of purity") that makes it a crime to brew beer with ingredients other than hops, malt, and water. In addition many of the popular beers are additive-free and beers from local microbreweries generally use only the finest natural ingredients. Read the labels and find a brand you like.

See also FOOD.

Blankets and Afghans

🠅 NATURAL FIBERS. Choose blankets and afghans made of natural fibers: cotton, silk, cotton-linen or cotton-wool blends, unmothproofed alpaca, and unmothproofed wool. Cotton thermal blankets can be found in department stores; other fibers can be ordered by mail. ***Chambers*** has some luxuriously beautiful blankets; ***Garnet Hill*** carries fine cotton and wool blankets.

See also TEXTILES.

Bleach

🠅 PREVENT MINERAL DEPOSITS AND SOAP SCUM FROM BUILDING UP ON FABRICS. Mineral deposits and soap scum cause fabrics to look dull and dingy. Use a WATER SOFTENER to prevent buildup and make the use of bleach unnecessary.

↑ ORGANICALLY GROWN NATURAL/RENEWABLE INGREDIENTS. Not yet available to my knowledge, but watch for them.

↑ NATURAL/RENEWABLE OR HYBRID-NATURAL INGREDIENTS. A few brands are available in natural food stores, greenstores, and green catalogs.

↑ HYDROGEN PEROXIDE. This is generally made from nonrenewable resources, but decomposes into hydrogen and oxygen. Safer than chlorine (hydrogen peroxide is often used as a topical antiseptic and mouthwash), it is used as a textile bleach in industry. Available in drugstores and chemical supply houses. Experiment with different concentrations in different amounts to find the level of whitening you need (start with a little at first—you can bleach something more, but once it's bleached, you can't undo it!).

↑ NONTOXIC/NONRENEWABLE INGREDIENTS. A few brands are available in natural food stores, greenstores, green catalogs, and hardware stores.

↓ TOXIC/NONRENEWABLE INGREDIENTS. Don't use products that contain sodium hypochlorite, lye, artificial dyes, detergents, fluorescent brighteners, or synthetic fragrances. Note warning on package label.

See also CLEANING PRODUCTS.

Bread

↑ ORGANICALLY GROWN OR BIODYNAMICALLY GROWN, UNBLEACHED, WHOLE-GRAIN. Easy to find in natural food stores and catalogs, but also try your local specialty bakeries—many are making very creative, delicious breads with organically grown grains. Natural food stores generally have whole-wheat and other whole-grain breads—in addition to regular loaves, try sprouted bread and unleavened breads. Bread made with biodynamically grown wheat can be ordered by mail from ***Nokomis Farms.***

↑ NATURAL. Most supermarket breads have some additives, although French bread is usually made from nothing more than flour, water, and starter, and comes unsliced in a paper bag. Most national brands are made with bleached white flour, although some have "whole-wheat" bread with some percentage of whole grain.

See also FOOD.

Brooms

↑ NATURAL BROOMCORN. Most brooms are made from natural materials and many broom handles are made from reclaimed scrap wood (whether so labeled or not). For a special broom, order the ***Berea College Student Craft Industries*** catalog—their fund-raising program employs students from rural Appalachia to make traditional local crafts. They have a wide variety of broom styles, handmade from natural broomcorn.

↓ PLASTIC BROOM BRISTLES. Made from nonrenewable resources, these are not biodegradable.

Butter and Oils

↑ ORGANICALLY GROWN, UNREFINED, PRESSED OILS. Natural food stores and catalogs carry a variety of such vegetable oils. They are more healthful to use because they are polyunsaturated fats. Unrefined "pressed" or "expeller pressed" oils are squeezed out using a mechanical process that does not use chemicals. Pressed olive oils are designated as "extra virgin" (from the first pressing) or "virgin" (from the next pressing). These oils retain colors, flavors, and aromas from their original sources, and can enhance or clash with the foods they are combined with.

↑ ORGANICALLY GROWN BUTTER. I've seen Wisconsin-made organically grown butter at my local natural food store and I can also buy butter made by a local organic dairy.

↑ NATURAL BUTTER. There are few additive-free butters—most are colored seasonally in order to maintain consistent color year-round. Dairies may use artificial colors or sometimes the hybrid-natural colors annatto and carotene (which may be, but are not always, preserved with BHA and BHT) without listing them on the label as ingredients. Find a local brand and check with the dairy about the colorants they use, or buy sweet (unsalted) butter, which usually does not contain colors (look in the freezer case at your supermarket or natural food store if it's not right next to salted butter).

↓ REFINED OILS. Most common oils found in supermarkets have been extracted using a petrochemical solvent. Then they are bleached, filtered, and deodorized, making the natural source indistinguishable. Preservatives are usually added. Read the labels, though, as there are a few oils that are preservative- and pesticide-free.

↓ MARGARINE. Hydrogenation (bubbling hydrogen gas in the presence of nickel through a tank of liquid polyunsaturated oil) is used to solidify cheap liquid vegetable fats, such as corn and safflower oil, into margarine. The process turns these polyunsaturated fats into saturated fats. So when you think you are getting a polyunsaturated fat in your margarine, you are actually getting the very saturated fats the margarine sellers claim they are helping you avoid. Better to eat butter in moderation, made simply by whipping cream, than such a technologically altered food as margarine.

See also FOOD.

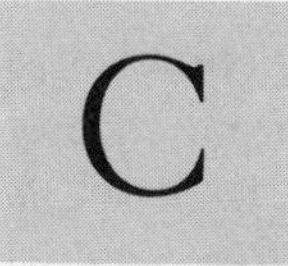

Candles

↑ BAYBERRY CANDLES. A New England tradition, bayberry candles are made from bayberries, which are boiled in water to release the wax. Look for them made by a local candlemaker, or order from ***Vermont Country Store.***

↑ BEESWAX CANDLES. These are sold in most candle shops and natural food stores. They are often bleached or tinted with artificial colors, so look for those that are their natural deep-honey color. ***Erwin's Bee Farm*** has a great selection of natural beeswax candles in many sizes and shapes, as well as beeswax for making your own.

Caulk

↑ NATURAL/RENEWABLE OR HYBRID-NATURAL INGREDIENTS. Natural caulks and joint sealants are not widely available, but are made by **Auro** and **Livos.**

↑ NONTOXIC/NONRENEWABLE INGREDIENTS. Nontoxic joint compound, caulking compound, and spackling compound is made by **AFM.**

↑ ROPE CAULK. Nontoxic and odorless, this can be used for many applications. Order by mail from ***Real Goods.***

♦ TOXIC/NONRENEWABLE INGREDIENTS. Other caulks are made from nonrenewable ingredients and contain toxic solvents.

Cereal

♦ ORGANICALLY GROWN WHOLE-GRAIN. Easy to find in natural food stores and catalogs. Several brands rival the supermarket array in variety of flavors—you'll find organically grown whole grains flaked and popped and sweetened and in granola.

♦ NATURAL. Only a few supermarket varieties are additive- or even sugar-free. Read labels carefully, as the presence of preservatives is often noted in a line separate from the ingredients because they are added to the package lining. At the very least buy a brand in a recycled paperboard box.

See also FOOD.

Cheese

♦ ORGANICALLY GROWN. A couple of national brands are available in natural food stores and catalogs—look for local brands that are made from organic milk. There are a few special cheeses you can order by mail.

Dutch Mill Cheese Shop carries cheese made by the Amish using milk from their own farms. While the Amish don't particularly claim their farms are organic, they use traditional farming methods which include horse plows and soil fertility plans (they pass their farmland from generation to generation, so have a practical interest in sustaining their land). Their cheeses are clotted with a vegetable enzyme and colored with vegetable color.

Shelburne Farms makes a farmhouse cheddar from the milk from their herd of Brown Swiss cows, descended from stock raised for cheesemaking in Swiss mountain villages. Their farm is an education center for demonstrating stewardship of natural and agricultural resources.

♦ NATURAL. Many brands and varieties of additive-free, natural cheeses are available at your supermarket, natural food store, and specialty gourmet and cheese shops. High-fat cheeses contain more pesticide residues because pesticides in milk get stored in the fat, so choose low-fat varieties.

➧ PROCESSED CHEESES, FLAVORED SPECIALTY CHEESES, CHEESES IN AEROSOL CANS. Avoid these as they are filled with additives and highly processed.

See also FOOD.

Cleaning Products

Cleaning products are the among the few household products regulated by the Consumer Product Safety Commission under the 1960 Federal Hazardous Substances Labeling Act. According to this law, cleaning products that are harmful to human health must carry various warnings on their labels.

If a cleaning product contains a chemical that is hazardous, it must by law specify the degree of toxicity by use of a signal word (DANGER: POISON, WARNING, or CAUTION). In addition it must state "the common or usual name or the chemical name . . . of the hazardous substance or of each component which contributes substantially to its hazard." Other labeling requirements include a statement telling users how to avoid the hazard (and, if necessary, safe use instructions), name and location of the manufacturer or distributor, instructions for handling and storage of packages that require special care, and a warning to keep out of the reach of children.

Manufacturers are required to keep records of all adverse health effects associated with the substance they produce. There is no federal law requiring premarket safety testing by the manufacturer, however, and the hazards of some products have sometimes not been revealed until after complaints were received by the Consumer Product Safety Commission.

The real safety or danger of cleaning products is difficult to assess because manufacturers are not required to list exact ingredients on the label. You can't look at a label to be sure, for instance, that a certain furniture polish doesn't contain nitrobenzene (a substance commonly used in furniture polish that could be fatal if swallowed), or that a mold and mildew cleaner is free from pentachlorophenol (another commonly used deadly substance); however, these ingredients should be listed on an MSDS. Some product ingredients, though, are protected by trade secrets and even the government and poison-control centers cannot find them out.

The best information we can get from poison-control centers is general lists of the chemicals commonly used in specific categories of products. Which brand-name products do or do not actually contain these substances is anybody's guess, unless the manufacturer voluntarily reveals the product's ingredients.

I believe that it is imperative to list all the product ingredients for cleaning products, particularly if they are toxic. Simply warning consumers that there is a danger does not give us enough information to make an informed choice. Industry has argued that consumers won't know what the words mean anyway, or that they will be alarmed if they know that the product contains certain toxic substances, but I say better informed than ignorant—we have a right to know what we're buying.

Philip Dickey of the Washington Toxics Coalition has suggested the following schemes (with examples) for more informative labeling on cleaning products. With labels such as these, we can know what each product is made from and find out for ourselves the safety or danger of the ingredients.

Ingredients only
Ingredients: water, sodium dodecylbenzene sulfonate, sodium lauryl sulfate, ethanol, 2-butoxy ethanol, triethanolamine, and methyl paraben.

Functions only
Ingredients: water, surfactants, grease cutter, buffering agent, preservative.

Ingredients and Functions
Ingredients: water, surfactants (sodium dodecylbenzene sulfonate, sodium lauryl sulfate), grease cutters (ethanol, 2-butoxy ethanol), buffering agent (triethanolamine), preservative (methyl paraben).

Ingredients, Percentages, and Functions (my personal favorite)

Ingredients:

Surfactants
sodium dodecylbenzene sulfonate 10%
sodium lauryl sulfate 5%

Grease Cutters
ethanol 15%
2-butoxy ethanol 3%

Buffering Agent
triethanolamine 10%

Preservative
methyl paraben 0.2%

Other
water 56.8%

Also required on the label by the Hazardous Substances Act are first-aid instructions, but don't rely on them in an emergency. A study done by the New York Poison Control Center found that 85 percent of product warning labels they studied were inadequate. Some labels list incorrect first-aid information, and others warn against dangers that don't even exist.

The environmental effects of cleaning products are not required on the label.

Green Seal has issued standards for household cleaners, though as of this writing none have been certified. Their standards require that the product not contain certain substances (too numerous to list here) that are toxic to humans or aquatic life and that each ingredient be biodegradable.

🠅 MAKE YOUR OWN. Very effective cleaning can be done with a few simple ingredients you probably already have around your kitchen: baking soda, lemon juice, and distilled white vinegar. Basic tips on how to use them for a particular purpose are given under the individual product entries. The comprehensive how-to natural cleaning guide to have on hand is *Clean & Green,* which gives approximately five hundred formulas for making your own cleaning products.

🠅 ORGANICALLY GROWN NATURAL/RENEWABLE INGREDIENTS. To my knowledge, there are no cleaning products currently available made with organically grown ingredients; however, it seems within the realm of possibility for the future as more organically grown ingredients become available at lower prices. Watch for them.

🠅 NATURAL/RENEWABLE OR HYBRID-NATURAL INGREDIENTS. Natural food stores now have full cleaning products departments—go to your local store and choose from a wide variety. Supermarkets also have a few natural cleaners, such as the old faithful **Bon Ami Polishing Cleanser** and the new **Heinz All Natural Cleaning Vinegar,** which is twice the strength of food-grade vinegar and comes in a spray bottle (for reasons unknown to me, it also contains natural lemon and pine flavor, lactic acid [from starch, milk, molasses, potatoes, or other natural sources] and polysorbate 80 [a surfactant emulsifier and dispersing agent made from sugar]—all natural ingredients, but why not just use plain vinegar?). Though some "natural" cleaning products are not made entirely from renewable resources—they may contain some minor ingredients made from nonrenewable resources—they generally list their ingredients, so you can look them up and find out their original source.

🠅 NONTOXIC/NONRENEWABLE INGREDIENTS. There are a number of new cleaning formulas that, while not made from renewable ingredients, are less harmful to health than standard cleaning products. Most stores that sell cleaning products now sell some brands labeled "nontoxic." These can generally be relied on to be relatively safe to use, but may have utilized toxic chemicals in their manufacture (one very popular "nontoxic" cleaner is actually made from a toxic chemical, diluted with water to the point where a toxic warning label is no longer required).

↓ TOXIC INGREDIENTS. These are easy to spot because they usually have warning labels and don't have the ingredients listed.

EarthRite Household Cleaners, Clear Magic Cleaners, and **Planet Cleaning Products** (often sold in hardware stores) are all certified biodegradable by Scientific Certification Systems.

See also AIR FRESHENERS AND ODOR REMOVERS; ALL-PURPOSE CLEANERS; BASIN, TUB, AND TILE CLEANERS; BLEACH; BROOMS; DISHWASHER "DETERGENT"; DISHWASHING LIQUID; DISINFECTANTS; DRAIN CLEANERS; FABRIC SOFTENERS; FURNITURE AND FLOOR POLISH; GLASS CLEANERS; LAUNDRY SOAPS AND DETERGENT; MOLD CLEANERS; OVEN CLEANERS; RUG, CARPET AND UPHOLSTERY SHAMPOO; SCOURING POWDERS; SHOE POLISH; SILVER POLISH; SOAP; and WATER SOFTENERS.

Clothes Dryer

↑ SUN AND AIR DRY. Get some clotheslines, wood clothespins, and wooden drying racks (sold in most greenstores and catalogs).

↑ ENERGY-EFFICIENT MODELS. You can save a significant amount of energy by purchasing a model that has a sensor to detect dryness and shuts off when the clothes are dry, rather than simply running for a selected period of time.

Clothes Washer

↑ ENERGY- AND WATER-SAVING MODELS. Saving water and saving energy go hand in hand because approximately 90 percent of the energy used by clothes washers goes to heat the water. Energy-saving features, then, include options that allow for cold-water rinse and multiple washload size settings. The less hot water you use, the more energy you'll save. Front-loading washing machines use one-third less water (and energy) than the standard top-loading model, and *Consumer Reports* says they have better overall washing performance. All new clothes washers must display EnergyGuide labels. Keep in mind that models with smaller capacities will have better energy ratings, but you may have to run them more often. Make your selection considering both the EnergyGuides and the energy-saving features. The *Consumer Guide to Home Energy Savings* lists the most efficient brands of clothes washers.

Clothing

🠝 Timeless Style. Instead of buying the latest fashions that need to be changed every year, choose classic styles that look good on you and make you feel attractive. Wear what you want to wear instead of what the fashion magazines tell you. Have your own individual style.

🠝 Organically Grown Cotton. Organic cotton clothing first became available commercially in 1991—among the pioneers were George Akers' Green Cotton Environment which is now **O wear** (a division of Vanity Fair), and EcoSport which is sold through the ***Seventh Generation*** catalog. In the beginning the only organically grown cotton clothing available was unbleached creamy white knit fabric, but now a number of companies are breaking new ground in more sustainable dye processes.

Another leader in more sustainable cotton clothing is Esprit with its **Ecollection** (available in stores and through their ***Ecollection*** catalog). The purpose of Ecollection is to act as a research laboratory to develop more environmentally sound ways to manufacture clothing and still be fashionable. They say "Our findings are neither conclusive or complete—Ecollection is an evolutionary process."

The first Ecollection debuted in 1992 using FoxFibre, organically grown cotton, low-impact dyes, unbleached fabrics, biodegradable enzyme washes, vegetable dye prints, nonelectroplated zippers, tagua nut buttons sustainably harvested from the rain forest, recycled glass buttons made by traditional handicraft workers in Ghana, and handpainted wooden buttons from a low-income cooperative in North Carolina. In their second season organically grown cotton, nonelectroplated zippers, and tagua nut buttons were incorporated into their main lines, and they introduced recycled wools and wools that come from black and gray sheep (so no dyes need be used for color and texture), and the cellulose-synthetic Tencel.

By all indications organically grown cotton is gaining in popularity and in the future will be widely available. In addition to the major companies, organic cotton clothing is available through smaller businesses. ***Organic Cottons*** has sportswear for the whole family at discount prices and ***Emel Organic Clothing*** sells original "more mainstream" designs for women.

🠝 "Green Cotton." A good source for green cotton clothing is the ***Seventh Generation*** catalog, with a fine selection of clothing for the whole family.

⬆ RECYCLED. Yarn and socks made from recycled cotton can be ordered from ***Seventh Generation*** and outdoor apparel made from DyerSport E.C.O., a recycled polyester fabric made from recycled PET plastic soda bottles, is available from ***Patagonia.***

Coffee

⬆ ORGANICALLY GROWN. This is easy to find in natural food stores and catalogs—even my local supermarket carries it now. Most organically grown coffees have the added social benefit of being produced by a workers' coop.

⬆ NATURAL. All coffees are natural in that they do not contain additives; however, they are often grown with pesticides so toxic they have been banned in the United States. If you prefer decaffeinated, choose steam- or water-processed varieties to avoid the hexane and methylene chloride used in the decaffeinating process.

⬇ FLAVORED INSTANT SPECIALTY COFFEES. These usually contain artificial additives.

See also FOOD.

Coffee Filters

⬆ FRENCH PRESS POT. A glass pot with a plunger that presses the coffee grounds to the bottom. Sold at specialty coffee stores.

⬆ REUSABLE GOLD MESH. This will last a lifetime.

⬆ COTTON CLOTH. Available in most natural food stores and catalogs.

⬆ UNBLEACHED PAPER. Easy to find where coffee filters are sold.

⬇ BLEACHED PAPER. Don't use, as these can release trace amounts of toxic dioxin into the coffee.

Comforters and Quilts

⬆ NATURAL FIBERS. Many cotton-covered down-filled comforters are available in a wide variety of styles and price ranges. Also available are cotton-filled, wool-filled, and silk-filled comforters. Cotton-filled quilts are making a renaissance too—I've seen them in import stores and department stores. Both ***Chambers*** and ***Garnet Hill*** have luxurious comforters; ***The Company Store*** has a good variety of styles and sizes at reasonable prices, and will also revitalize your old comforter or custom-make one for you.

🠉 ORGANICALLY GROWN. ***Jantz Design*** and ***Heart of Vermont*** both make comforters with organically grown materials.

See also TEXTILES.

Computers

🠉 ENERGY-EFFICIENT. Computers now represent 5 percent of America's commercial energy consumption, and that doesn't include home computers. Virtually all major computer manufacturers are at work developing energy-efficient desktop models. Leading the way is the IBM **Energy Saver** which cuts power by three quarters compared to a conventional system by borrowing technology from laptops: flat-panel LCD screen, automatic sleep mode when the computer is on but idle, and lower voltages. In addition some of the new "green PCs" will have chips made by processes using less harmful or recycled chemicals (CFCs are being phased out), and housings made of recyclable plastic.

🠉 RECYCLE RIBBONS, BATTERIES, AND LASER CARTRIDGES. Here in the San Francisco Bay Area we have a number of places that will collect and refill ribbons and laser cartridges (ask at your computer store). Major manufacturers of toner cartridges are willing to take them back for recycling, so if you can't find a local place, send it back to the manufacturer.

🠉 SAVE PAPER. Preview your documents electronically before you print them out, save unusable printouts and use the backs to print drafts or as scratch paper, and print only when necessary. Take advantage of computer capabilities (such as sending faxes directly from the computer instead of printing them out first, or sending documents via modem) to reduce paper needs.

The Green PC lists one hundred practical things you can do to reduce the environmental hazards of computing.

Condiments

These include barbecue sauce, chutney, ketchup, mayonnaise, mustard, olives, pickles, relish, salad dressing, salsa, soy sauce, and vinegar.

🠅 ORGANIC. There are many condiments made from organically grown ingredients sold in natural food stores. Some will be primarily organic (such as a soy sauce made from organically grown soy beans) and others will have a few minor organic seasonings (such as organically grown herbs or garlic in a salad dressing). Either way, your purchase of these products helps support the growth of sustainable agriculture.

🠅 NATURAL. All condiments sold in natural food stores are additive-free, but many supermarket brands contain additives, so check labels carefully.

See also FOOD.

Cooking Appliances

🠅 SOLAR COOKERS. Use the power of the sun to slow-cook or bake. After about an hour of preheating, these cookers generally need no more working time than conventional ovens. A ready-made sun oven can be ordered from ***Real Goods.***

To Cook a Casserole

APPLIANCE	TEMPERATURE	TIME	ENERGY	COST
Electric oven	350°F	1 hour	2.0 kWh	16¢
Convection oven	325°	45 min.	1.39 kWh	11¢
Gas oven	350°	1 hour	0.112 therm	7¢
Frying pan	420°	1 hour	0.9 kWh	7¢
Toaster oven	425°	50 min.	0.95 kWh	8¢
Crockpot	200°	7 hours	0.7 kWh	6¢
Microwave oven	High	15 min.	0.36 kWh	3¢

Source: *Consumer Guide to Home Energy Savings.*

🠅 ENERGY-EFFICIENT COOKING. Different kinds of appliances use different amounts of energy.

The least energy-efficient is to use an electric oven or convection oven. Gas ovens, cooking in a frying pan on the top of a stove, toaster ovens, and Crockpots are in a similar, fairly efficient range, and microwave ovens are the most energy-efficient.

Even though they are the most energy-efficient, microwave ovens leave something to be desired in the way food looks and tastes, and they have potential health dangers. If you use a microwave oven, you should be careful of leakage as even microwave ovens that are functioning perfectly can emit microwaves. FDA safety standard limits allow microwave emission of up to 1 milliwatt per square centimeter (1 mW/cm^2) when the oven is purchased, and up to 5 mW/cm^2 after the oven has been in use. Animal studies done by Bell Telephone Laboratories have convinced this company to establish tighter safety regulations, recommending that daily exposure not exceed 1 milliwatt. Illness associated with microwave exposure includes headaches, fatigue, irritability, sleep disturbances, weakness, and heart and thyroid problems.

Consumer Reports magazine suggests that you can minimize your risk by keeping a reasonable distance from the oven while it is in operation (the farther, the better), and try to operate and maintain the oven in such a way that will minimize leakage. Make sure the oven door closes properly and that no damage occurs to the hinges, latches, sealing surfaces, or the door itself. Make sure that no soil or food residue accumulates around the door seal, and avoid placing any objects between the sealing surfaces. You can also test the oven for leakage with a handheld detector.

Gas is a popular choice for cooking because it is economical to purchase, burns efficiently and more cleanly, and gives the cook a greater level of control. Choose a stove with an electronic ignition instead of a pilot light; you'll save more than 30 percent of gas use. One disadvantage to using gas are combustion by-products, particularly carbon monoxide. In California by law we have received warnings that natural gas contains benzene, which is known to cause cancer. So when using a gas appliance of any kind it's important to watch out for gas leaks and to adequately vent combustion by-products. EnergyGuide labels are not required on cooking appliances.

Cookware

♠ CLAY. Clay cookers are made from the abundant clay of the earth, use little energy while cooking, and can be broken up and returned to the earth at the end of their useful life. Clay vessels have been used for cooking for millennia, and are still the most sustainable. You soak them in water before cooking, and the food gently steams as the hot moisture comes out of the clay. I have a wonderful, handmade domed terra-cotta cooker made by African women, but if you can't find (or have made) a handmade one, you can buy machine-made clay cookers (such as **Römertopf**) at most major department stores and cookware stores, or in the ***Williams-Sonoma*** catalog. I've also seen clay bread pans and muffin pans at craft fairs, so look around.

🡅 GLASS. Though the glass used to make cookware does not have recycled content, nor is it recyclable (it has additives to make it heat-resistant that are incompatible with recycled bottles), it is still made from abundant minerals by less environmentally damaging processes than metals. Glass cookware and bake-ware can be purchased everywhere cookware is sold.

🡅 🡇 STAINLESS STEEL. I'm having a hard time figuring out what to recommend about stainless steel cookware. It has generally been considered the best choice for cooking because it is sanitary, nonporous, and the metals were thought to be highly stable. But a 1993 study has shown that stainless steel cookware can add nickel to food, a toxic metal that has been shown to cause cancer in humans when inhaled. Researchers at Pennsylvania State University in Erie tested seven brands of stainless steel pots by using them to boil a mixture based on the acid content of fruits, vegetables, and starchy foods. Nickel was found to be present in each. Environmentally, too, the mining and manufacture of steel is a highly technological, energy-intensive, and polluting process. On the other hand, the most energy-efficient cookware I know of is made from stainless steel. **Durotherm Plus** (order through ***SelfCare Catalog***) has double-walled sides and insulated lids which allow you to slow cook at low energy-saving temperatures. Because the pots retain heat, foods will continue to cook even after they are removed from the heat, saving even more energy. And pressure cookers—which many people use to cook healthful whole grains and beans in energy-saving minutes instead of hours—are also made from stainless steel. I haven't given up my stainless steel yet, but I'm thinking about it. Carbon steel does not have nickel as an ingredient and therefore doesn't release it. Woks are made from carbon steel and heavy carbon steel sauté pans are available at restaurant supply houses. Weigh the pros and cons and do what is right for you.

🡅 CAST IRON. This has been the mainstay cookware for generations. Durable, inexpensive, and, though made from metal, it is simple in material and processing.

🡅 ANODIZED ALUMINUM. Aluminum is a preferred metal for cookware because it distributes heat evenly. Though there are health problems associated with aluminum cookware (see below), if a label says the cookware is made from "anodized" aluminum, it is safe to use. This means that the aluminum was dipped into a hot acid bath that seals the aluminum by changing its molecular structure. Once anodized, the aluminum will not leach into food. If you are considering buying aluminum cookware, call the manufacturer and see if it has some recycled content.

🠗 ALUMINUM. Aluminum salts can leach from the pot into the food being cooked, particularly if it is acidic, causing a number of unpleasant symptoms. Most aluminum cookware manufactured today is anodized (see above). Check the label carefully and watch out for nonanodized aluminum if you buy used cookware at flea markets or thrift shops. The sale of aluminum-lined cookware is prohibited in Germany, France, Belgium, Great Britain, Switzerland, Hungary, and Brazil, but is still permitted in America.

There are some brands of cookware that use aluminum for the base of the pan and then line the pan with stainless steel or some other finish. Cookware containing aluminum is safe to use only when the aluminum does not come in contact with the food.

🠗 COPPER. Copper heats quickly and has very even heat distribution, but, like aluminum, it can leach into food. Only buy copper cookware if the copper is on the outside of the pan, and does not come into contact with food.

🠗 NO-STICK AND PORCELAIN CERAMIC FINISHES. Besides being made from a nonrenewable resource, some of these finishes chip and scratch easily and can contaminate your food with bits of plastic or ceramic while you are cooking. If you want a no-stick pan, get a cast-iron pan and season it. Before using the pan, cover the bottom with cooking oil and place it on a warm burner for one hour. Wipe out the excess oil, leaving a thin film of oil on the pan. When you are cooking in any kind of pan, if you heat the pan first, then add your oil to the hot pan, and let the oil warm up before you put in the food, it won't stick.

Cosmetics

🠕 ORGANICALLY GROWN INGREDIENTS. Check the labels at your natural food store.

🠕 NATURAL/RENEWABLE AND HYBRID-NATURAL INGREDIENTS. Nowadays natural food stores are full of natural cosmetics that rival the department store brands in quality and appearance. The most natural and simplest that I know of are **Nature's Colors** (in some natural food stores and by mail). My favorite natural food store lipstick is made by **Kiss My Face**—it comes in beautiful natural colors (most "natural" lipsticks still use artificial colors) in little cases made from sustainably harvested wood.

See also BEAUTY AND HYGIENE.

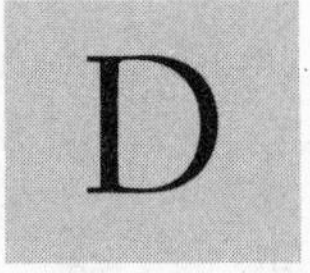

Dishware and Glassware

↑ RECYCLED GLASS. A variety of dish and glass styles made of recycled glass can be purchased at most import stores, greenstores, and mail-order catalogs. Often these are handmade by recycling old soda bottles, and can be recycled along with glass bottles and jars.

↑ POTTERY WITH RECYCLED, LEAD-FREE GLAZE. Glaze that runs off during firing is collected and periodically recycled. Some styles of *Health Ceramics* (sold in gift shops and by mail) have recycled glaze.

↑ POTTERY FINISHED WITH A LEAD-FREE GLAZE. Local potters are a good source of unusual and imaginative designs. Potters often label their pieces "lead-free" if a lead-free glaze is used—if there isn't a label, ask the potter, as lead-free glazes are becoming almost standard among craftspeople concerned about their own safety. Pottery and most glazes are made from naturally occurring mineral clays (not renewable, but abundant, used straight from the earth). Broken pottery can be ground up and used again to make new pots.

↑ CLEAR OR COLORED GLASS. Glass is made from silica, silicates, and other minerals (not renewable, but silica is the most abundant mineral on earth). Plain glass glassware is sold in every store that carries glassware; clear glass dishware is sold in places like Crate & Barrel and import stores. Made from mined minerals, it cannot be recycled.

↓ STANDARD CHINAWARE. The federal government prohibits the sale of dinnerware that releases lead in amounts greater than 2,000 ppb (which prevents direct cases of lead poisoning), but the state of California requires warning labels on any dishware that releases lead in amounts greater than 224 ppb to protect against long-term health risks. Most major manufacturers of dinnerware sold in department stores and home decorating shops still use lead glazes, without labeling them as such. If you want to buy this kind of dinnerware, ask the salesperson and verify with the manufacturer whether or not a lead-free glaze was used on the particular style you are interested in. Generally patterns with bright colors contain lead; white and subdued earth tones do not.

🡇 IMPORTED POTTERY WITH BRIGHT GLAZES. The biggest offenders of lead in dishware and glassware are the brightly colored pottery pieces from foreign countries. If in doubt don't buy it.

🡇 LEAD CRYSTAL. Lead is added to the basic glass formula to make the glass sparkle more by refracting light. In some states lead crystal now requires a health warning label.

Dishwasher "Detergent"

🡅 ORGANICALLY GROWN NATURAL/RENEWABLE INGREDIENTS. Not yet available to my knowledge, but watch for them.

🡅 NATURAL/RENEWABLE OR HYBRID-NATURAL INGREDIENTS. A few brands are available in natural food stores, greenstores, and green catalogs. A soap-based compound can be ordered by mail from ***Cal Ben Soap Company.***

🡅 NONTOXIC/NONRENEWABLE INGREDIENTS. A few brands are available in natural food stores, greenstores, green catalogs, and hardware stores.

🡇 TOXIC/NONRENEWABLE INGREDIENTS. Don't use products that contain chlorine, artificial dyes and fragrances, or detergents. Note warning on package label.

See also CLEANING PRODUCTS.

Dishwashers

🡅 WASH YOUR DISHES BY HAND. Regardless of how water- and energy-efficient a dishwasher might be, nothing is more efficient than washing dishes by hand if you don't leave the water running.

🡅 ENERGY-SAVING MODELS. About 80 percent of the total energy used by dishwashers goes to heat the water; therefore models that use less water use less energy (and, of course, save water, too). Most dishwashers use between eight and fourteen gallons of water for a complete wash cycle, so choosing the right model can save almost half the energy needed. EnergyGuide labels must be displayed on all new dishwashers sold. The *Consumer Guide to Home Energy Savings* lists the most efficient brands and gives tips for choosing the most efficient dishwasher for your needs.

Dishwashing Liquid

🠅 PLAIN LIQUID SOAP.

🠅 ORGANICALLY GROWN NATURAL/RENEWABLE INGREDIENTS. Not yet available to my knowledge, but watch for them.

🠅 NATURAL/RENEWABLE OR HYBRID-NATURAL INGREDIENTS. Several brands are available in natural food stores, greenstores, and green catalogs. A pleasant almond-scented soap-based product can be ordered by mail from ***Cal Ben Soap Company.***

🠅 NONTOXIC/NONRENEWABLE INGREDIENTS. Several brands are available in natural food stores, greenstores, and green catalogs.

🠇 TOXIC/NONRENEWABLE INGREDIENTS. Don't use products that contain detergents, or artificial dyes or fragrances.

See also CLEANING PRODUCTS.

Disinfectants

🠅 BOILING WATER. To sterilize (kill all germs present), immerse in boiling water.

🠅 CLEAN REGULARLY WITH PLAIN SOAP AND WATER. Even just a rinse of hot water will kill some bacteria.

🠅 KEEP THINGS DRY. Bacteria, mildew, and mold cannot live without dampness.

🠅 NATURAL/RENEWABLE, HYBRID-NATURAL INGREDIENTS, NONTOXIC/NONRENEWABLE INGREDIENTS. None available to my knowledge, but watch for them in natural food stores, greenstores, and green catalogs.

🠇 TOXIC/NONRENEWABLE INGREDIENTS. Virtually all disinfectants on the market kill germs with such toxic chemicals as cresol, phenol, formaldehyde, ammonia, and chlorine. Note warning on package label. These will "kill germs on contact," but will not kill all the germs present.

See also CLEANING PRODUCTS.

Drain Cleaners

⬆ PREVENT CLOGGED DRAINS. Use a drain strainer to trap food particles or hair that might cause a clog. Don't pour grease down the drain (dump it into the garbage or into a grease can to be reused instead).

⬆ MECHANICAL MEANS. Use an old-fashioned plunger. There's also a nifty product called the **WorldWise Easy-Clear Sink Trap** that replaces the curved part of your pipe under the sink and opens at the curve so you can pull the clog out. And if the clog is further down the pipe, you can use **Drain King,** a device that creates water pressure with water from your garden hose to push the clog through. If worse comes to worst, use a mechanical snake or call a professional to bring in mechanical equipment. Mechanical devices are sold in hardware and home improvement stores.

⬆ BIOLOGICAL MEANS. These products are designed to remove soap, hair, grease, and other organic materials that coat the entire length of pipe and cause slow drains. Enzyme action removes this buildup so pipes can flow freely. One such product, **Drain Care** (sold in hardware stores), is certified biodegradable by Scientific Certification Systems.

⬆ HYDROGEN PEROXIDE. Pour one quarter cup 3 percent hydrogen peroxide down the drain. Wait a few minutes, then plunge. Repeat a second time if needed. This has been known to open clogged drains that have defied other methods.

⬇ TOXIC LYE-BASED CLEANERS. Do not use. Drain cleaners are among the most toxic consumer products sold, and don't work as well as mechanical means to clear drains. Note "POISON" warning on package label.

See also CLEANING PRODUCTS.

Dried Fruits

⬆ ORGANICALLY GROWN, UNSULFURED, SUN-DRIED. These are the best dried fruits you will ever taste. Absolutely free from chemicals, sun-dried fruits dehydrate slowly, preserving both the color and the flavor. If your natural food store doesn't carry sun-dried fruits, you can order them by mail from ***Everything Under the Sun*** or ***Walnut Acres.***

🠉 ORGANICALLY GROWN, UNSULFURED. Fruits of all kinds are commonly available in bulk in natural food stores and catalogs. Organically grown fruits are generally unsulfured, so they are not as brightly colored as the sulfured fruits.

🠋 SULFURED.Most dried fruits sold are sulfured; the label will say "contains sulfites."

See also FOOD.

Eggs

🠉 ORGANICALLY GROWN, FREE-RANGE. Easy to find in natural food stores, at farmers' markets, and in some supermarkets. Carton should give some indication of the diet and handling, even if it doesn't say "organic." Fertile eggs are often grown by organic methods, but the word *fertile* alone on the carton simply means that the eggs can be hatched into baby chicks; it does not mean that the eggs are organically grown unless there is some other indication. Free-range means that the hens have been allowed to run instead of keeping them confined; usually these types of growers will also use other organic methods, but, again, check the label.

🠋 STANDARD. Nonorganic eggs generally contain residues of pesticides, antibiotics, hormones, stimulants, tranquilizers, and fumigants.

See also FOOD.

Energy, Renewable

To live sustainably we will, within the foreseeable future, need to move from using nonrenewable fossil fuels as our source of power, to using renewable sources of energy. While solar, wind, hydroelectric and biomass already have limited installations as part of progressive utility companies, solar (and, to a lesser extent, wind power) is accessible to almost any homeowner (depending, of course, on the amount of sun at the specific site).

Though renewable energy equipment is not yet price-competitive in closely populated areas that are "on-the-grid," installations in remote areas (where it can be very expensive or even impossible to run power lines) have shown that renewable energy sources can provide all the energy needed for a private home.

Renewable energy is most workable on a small scale, where it is generated close to the point at which it is used. Therefore it is most appropriate for a single house or group of houses, or a community. It's something a group of individuals could do for themselves, instead of relying on their local utility to make the change.

There are a number of good catalogs that have been around for a while and have good information, good equipment, and technical assistance. ***Real Goods*** is the largest, with several catalogs and every piece of equipment and every solar toy and accessory you could imagine. ***Backwoods Solar Electric Systems*** is at the other end of the spectrum—a small home-based business that operates on solar energy with wind power as backup on stormy days when there is no sun. They have good background information that is easy for the beginner to understand. ***Integral Energy Systems, Jade Mountain,*** and ***Photocomm*** all have the basics, plus background information (other suppliers regularly advertise in magazines like the *Mother Earth News*). Order some catalogs and some books, and start exploring the possibilities appropriate for you.

Fabric

🠉 NATURAL FIBERS. Better fabric stores will carry a large variety of natural fiber fabrics.

🠉 ORGANICALLY GROWN COTTON. Organic cottons are now being used by individual weavers to make fabrics by hand. One such weaver is Ellen Anthony, who has her own small mail-order business ***Ocarina Textiles.*** She uses yarns that are undyed, unbleached, unmercerized, and organic when she can get them, and looms built early in this century, to weave upholstery fabric which she purchased from a defunct mill. She beautifully combines the natural creamy shade of cotton with the **FoxFibre** yarns in classic American domestic patterns. Another source is ***Fiber Naturals,*** which sells machine-woven unbleached organic cotton yardage in a variety of fabric types, along with **FoxFibre** fabrics. **Foxfibre Colorganics** fabrics can be ordered from the grower via its sister company ***Vreseis.***

See also TEXTILES.

Fabric Softeners

🡅 WEAR NATURAL FIBERS. Fabric softeners are formulated to reduce static cling in synthetic fabrics and are unnecessary with natural fibers.

🡅 MAKE YOUR OWN. To make natural fibers softer, pour one cup white vinegar into the final rinse water.

🡅 ORGANICALLY GROWN NATURAL/RENEWABLE INGREDIENTS. Not yet available to my knowledge, but watch for them.

🡅 NATURAL/RENEWABLE OR HYBRID-NATURAL INGREDIENTS. A few brands are available in natural food stores, greenstores, and green catalogs.

🡅 NONTOXIC/NONRENEWABLE INGREDIENTS. Choose an unscented sheet variety that goes into the dryer (available in every supermarket) over a liquid in a plastic jug or an aerosol spray applied to dry clothes.

See also CLEANING PRODUCTS.

Faucets

🡅 LOW-FLOW AERATORS. These can be attached to almost any fixture and will reduce the water flow to three gallons per minute. Some have on-off levers that allow you to restrict the flow without having to turn the water off once you've adjusted it to the correct temperature. Low-flow faucet aerators work by reducing water flow and increasing pressure, then mixing air with the water as it comes from the tap. Even though you're using a lot less water, it will seem like the flow is stronger. Most hardware and plumbing stores carry them, priced at less than four dollars.

🡅 LOW-FLOW FAUCETS. Replacing the faucet altogether is more expensive than buying an aerator, but new faucets flow at 2.75 gpm or less. When looking at new faucets, ask if there are any made without lead solder (this new exposure to lead has recently come to light, so it might be a while before lead-free faucets are sold). Green Seal has a certification standard of 2 gpm for lavatory faucets and 2.5 gpm for kitchen faucets. They also require that any contamination from toxic substances such as lead or chromium be less than levels allowed in the national drinking water standards.

Feminine Protection

⬆ Reusable Cotton Pads. This is the number-one most sustainable choice. ***New Cycle*** (in natural food stores and by mail) sells pads made from cotton flannel in feminine colors and patterns, and from naturally brown organically grown cotton knit. Or you can sew them up yourself from old flannel shirts and sheets.

⬇ Disposable Tampons and Pads. Most disposable "paper" pads and tampons are made with superabsorbant synthetic fibers and plastic. Synthetic fibers appear to increase the risk of toxic shock syndrome. In addition, disposable pads (and tampons, too, if they can't be flushed) end up in landfills.

Fire Extinguishers

⬆ CFC-Free Dry Chemical Type. These are made from sodium bicarbonate and monammonium phosphate (**First Alert** is one brand commonly available at hardware stores). The label might not say CFC-free—look for the words *dry chemical* somewhere in the promotional text on the label.

⬇ Contains CFCs. Fire extinguishers are the only product listed by the EPA as containing ozone-depleting halons. Do not use them.

First Aid

⬆ Home Remedies. Get a good book of home remedies to keep in your medicine cabinet. Often these simple ways are just what is needed.

⬆ Organically Grown Herbs. Herbs are the oldest form of medicine as well as the basis of modern pharmacology. Herbs and medicinal plants can be taken in many forms, including tinctures (the most concentrated, usually in an alcohol base), teas, pills, and pressed juices for internal use, and ointments, salves, and shampoos for external use. Your natural food store should have a whole array of herbal remedies and potions.

⬆ Homeopathic Remedies. Homeopathy is a medicinal system that stimulates our own innate healing and immune processes by using the energetic essences of plant, mineral, and animal substances captured by a water-based pharmaceutical process. Almost every pharmacy in France carries homeopathic medicines, and in Germany sixteen thousand physicians use this medically recognized system. Natural food stores in America generally carry homeopathic remedies, which usually come in a pill form that you dissolve under your tongue.

See also BEAUTY AND HYGIENE.

Fish and Seafood

⬆ DEEP-WATER. Select species that spend most of their lives in deep water, far out at sea and away from human pollution: herring, sardines, anchovies, small salmon (pink, coho, sockeye, and Atlantic), scrod, hake, haddock, pollock, mackerel, pompano, red and yellowtail snapper, striped bass, butterfish, squid, octopus, and tilefish. If you eat fish as a regular part of your diet, find a local fishmarket and get acquainted. Ask where the fish was caught and if it was treated with any preservatives. You might need to do a little digging, but you can probably find out the fishing practices and see if the fish are being sustainably harvested from the area.

⬇ COASTAL AND FRESHWATER FISH, SHELLFISH. Be careful with these species, as they are often contaminated by polluted waters. Fish can concentrate pollutants to levels up to two thousand times greater than found in the surrounding waters. Find out where fish is caught in your local area and the level of water pollution. If the water is relatively unpolluted, there's no reason not to eat the fish, especially if you catch them yourself.

See also FOOD.

Flour

⬆ ORGANICALLY GROWN, UNBLEACHED, WHOLE-GRAIN. These are easy to find in natural food stores and catalogs. If you bake regularly, you might want to invest in your own grain grinder, as freshly ground whole-grain flours have more nutrition and flavor.

⬆ UNBLEACHED. This is usually available at supermarkets. While it still contains pesticide residues, it is less processed than other white flour, and the omission of the bleaching makes little difference in baking.

See also FOOD.

Food

⬆ ORGANICALLY GROWN OR BIODYNAMICALLY GROWN FRESH, WHOLE FOOD, IN SEASON, AND LOCALLY GROWN. The most sustainable cuisine is created from whole fresh foods, in season for your locality, grown as close as possible to the place where you eat it.

In addition to growing your own or finding organically grown food at natural food stores and some supermarkets, roadside stands and farmers markets are another source. In some areas, farmers publish cooperative maps so you can pick your own produce. Join a Community Supported Agriculture (CSA) program. In a CSA members buy yearly subscriptions in a local organic farm and receive a weekly basket of whatever is being harvested.

If you have no local source, almost every kind of organically grown food there is can be ordered by mail. *Green Groceries* gives hundreds of mail-order sources for organic foods. Some good general catalogs to start with are ***Gold Mine Natural Food Company, Jaffe Brothers,*** and ***Walnut Acres.***

The few biodynamically grown foods available in this country are sold in natural food stores, through CSAs, and by mail.

⬆ ORGANICALLY GROWN PROCESSED FOODS. Many processed foods are now made with organic ingredients (natural food stores are full of them). While it's more sustainable to prepare food from scratch at home (less packaging and shipping and more nutrition), if you're going to buy potato chips or mustard anyway, buy a brand made with organically grown ingredients.

⬆ NATURAL FOODS. These are generally understood to be any packaged food product that does not contain any added "artificial" (having no counterpart in nature) additives, but it does not necessarily extend to the exclusion of pesticides or other chemicals used in processing.

Back in the early 1980s, in response to an explosion of food products claiming to be "natural," the Federal Trade Commission recommended standards for food products labeled "natural." At the time it was recommended that a definition be approved by Congress so that it could be enforced, but Congress temporarily tabled the issue, and it never came up again for review. In general, though, natural foods meet this definition: A food may be called natural only if it contains no artificial ingredients, and had no more processing than it would normally receive in a household kitchen . . . Minimal processing will include washing or peeling fruits or vegetables; homogenizing milk; freezing, canning, or bottling foods; grinding nuts; baking bread; and aging meats.

Additives (artificial and natural) are those ingredients added to the products above and beyond the main food ingredients. Additives in food are nothing new—salt was used as a preservative for raw meat centuries ago. But modern chemistry has given us many new, unnatural, and often toxic ways to preserve or alter food. We now have more than 8,600 food additives approved for use, of varying safety and toxicity.

Food additives are regulated by the Food and Drug Administration (FDA) under the Food Additives Amendment to the Pure Food and Drug Act, passed in 1958. Exempt from regulation are those additives Generally Recognized As Safe (GRAS). The list of GRAS substances was compiled by the FDA in the late 1950s by making a list of additives they assumed to be safe, and sending out the list with a questionnaire to about nine hundred scientists. Of the one third who responded, only about one hundred offered substantive comments. The GRAS list was finalized from these comments. Ten years later the FDA began to review the scientific literature to determine the safety of GRAS substances.

The Food Additives Amendment also contains the Delany Clause, which states that "no additive shall be deemed safe if it is found to induce cancer when ingested by man or animal."

Because the names of many additives do not identify whether their source is natural or synthetic, it is often impossible to know from reading a label if a food contains artificial additives. In general if the label says "artificial," or if you don't recognize an ingredient as a food, or if it doesn't say "natural" (as in "natural flavor" or "natural color"), you probably shouldn't buy it. But not all food additives are artificial or harmful. Many additives with strange-sounding names come from natural sources and are perfectly safe to use. To find out the derivation and safety of almost any food additive, see *A Consumer Dictionary of Food Additives*. Some natural/renewable, safe food additives that are often used in natural foods include:

Acetic acid—another name for plain vinegar

Agar-agar—derived from seaweed

Albumin—derived from egg whites

Alginates—derived from seaweed

Annatto—taken from a rain-forest tree

Beta-carotene—Vitamin A derived from carrots

Carrageenan—derived from seaweed

Citric acid—Vitamin C from citrus fruits

Dextrose—corn sugar

Dulse—derived from seaweed

Guar gum—made by grinding the seed of a plant cultivated in India

Gum tragacanth—the dried gummy exudate from a Middle Eastern plant

Lactic acid—made by fermentation of whey, cornstarch, potatoes, and molasses

Lecithin—isolated from eggs, soybeans, and corn

Pectin—taken from lemon or orange rinds

Sodium chloride—common table salt.

The presence of many other additives in food products may, however, be legally concealed. Processed food may use ingredients already containing additives without being required to list these additives on the label. Ham included in processed foods, for instance, will contain nitrates and nitrites, and the shortening used may contain BHA and BHT, but only "ham" or "shortening" will appear on the product label.

Your local natural food store is filled with natural packaged foods, and you won't have to read the labels to make sure they are free of artificial additives because any natural food store with a good reputation wouldn't even consider carrying such a thing (nor would their customers allow it). However, if you shop carefully, you'll also find many additive-free foods at your supermarket. Many major manufacturers have been making additive-free foods for years, and these are available at a lower cost than some natural brands with identical ingredients. (On the other hand, some supermarkets with natural food departments sell the same brands found in natural food stores at greatly inflated prices.)

➔ PROCESSED FOODS WITH ARTIFICIAL ADDITIVES. Most artificial additives are made from derivatives of crude oil, though a few come from renewable sources. Regardless of the source of raw materials, they are chemically altered to produce a synthetic substance. While they are not exceptionally harmful to health or to be avoided at all costs, there are negative health effects associated with them (particularly if they are ingested on a regular basis) and are without exception used in foods grown with pesticides and artificial fertilizers. Some artificial additives to watch out for are:

Artificial colors (also listed on labels as FD&C colors, or as a specific color such as FD&C Yellow #5)—made from coal tar. Shown to cause cancer in laboratory animals.

Artificial flavors—made from coal tar. Linked to hyperactive behavior in children.

BHA and BHT—made from coal tar. Linked to hyperactive behavior in children. On the GRAS list (Generally Recognized As Safe).

EDTA (ethylenediamine tetraacetic acid)—made from minerals. Irritating to the skin and can cause allergic reactions. Can cause kidney damage. On the FDA list for further study.

MSG (monosodium glutamate)—derived from glutamic acid, an amino acid. Animal studies show brain damage, stunted skeletal development, female sterility, and obesity. On the FDA list of additives that need further study for mutagenic and reproductive effects.

Nitrates and Nitrites—made from mineral salts. Can form carcinogenic nitrosamines when combined with amines in food, known to cause cancer in test animals. Added to most pork, especially processed meats, as well as some other meat, poultry, fish, and cheese.

Sulfites (sulfur dioxide, sodium sulfite, sodium and potassium bisulfite, sodium and potassium metabisulfite)—made from sulfur and mineral salts. Can cause severe allergic reactions, including breathing difficulties, gastrointestinal disorders, unconsciousness, hives, and anaphylactic shock. Found in many processed foods and almost always in wine and beer.

Federal food labeling laws require most food products to have nearly all their food additives listed on the label, thereby making them easy to spot and avoid. Any synthetic preservative must be specifically identified, and artificial colors and flavors must either be specifically named or described as artificial (except that artificial colors need not be mentioned on butter, cheese, and ice cream labels, or on unprocessed fresh foods such as eggs and oranges).

➧ IRRADIATED FOODS. Food irradiation is done by zapping food with a dose of radiation almost 60 million times that of a chest X ray. Gamma rays (radioactive by-products of the nuclear industry) or X rays are used to kill insects and bacteria, prevent sprouting, and slow rotting. While the process does not make the food itself radioactive, the chemical structure of the food is altered and there are a number of animal studies that show negative health effects. Some researchers suspect that a regular diet of irradiated food may cause leukemia, other forms of cancer, and kidney disease. As with some other products that initially appeared to be safe, such as cigarette smoking, pesticides, asbestos, and CFCs, I believe we may not learn the dangers of irradiated foods until they become widely used.

The facilities where irradiation takes place also pose health and environmental effects to workers and the general population. Besides potential problems with the transportation of radioactive materials, there have already been several radiation leaks and contaminations, as well as numerous safety violations in existing plants.

The FDA claims irradiation of food is safe and has already authorized the irradiation of fruits, vegetables, pork, chicken, herbs, spices, teas, and seeds. So far, spices are the most widely irradiated food in the United States. The FDA requires whole foods that have been irradiated to display on the label an international logo of a flower in a circle, but this is not a warning label, nor is any

wording required to indicate the product is irradiated. However, when used as an ingredient in processed foods or in spice blends, the fact that a food has been irradiated does not have to be indicated on the label. Irradiated foods can also be used in restaurant foods, school cafeteria foods, or institutional foods without notification.

The only precaution you can take against irradiated foods is to look for the irradiated logo on the label. I have never actually seen one on a food product, but by law it should appear on whole foods that have been irradiated. Because of food labeling laws that allow irradiated ingredients in processed foods without mentioning this on the label, we can assume that some irradiated ingredients may be present in processed foods without our knowledge.

Foods that are organically grown have not irradiated.

See also BEER; BEANS; BREAD; BUTTER AND OILS; CEREAL; CHEESE; COFFEE; DRIED FRUITS; EGGS; FISH AND SEAFOOD; FLOUR; GRAINS; HERBS AND SPICES; JAMS, JELLIES, PRESERVES, AND FRUIT SPREADS; JUICE; MEAT AND POULTRY; MILK; NUTRITIONAL SUPPLEMENTS; NUTS, SEEDS, AND BUTTERS; PASTA AND SAUCE; SWEETENERS; TEA; WINE; and YOGURT.

Food Storage

♦ GLASS JARS. Decorative glass jars are sold in natural food stores and import shops, or you can save and reuse jars from food purchases.

♦ COTTON BAGS. These are handy for storing grains, potatoes, onions, produce, and bread. The best way to keep lettuce in the refrigerator is to wash it leaf by leaf, shake off the water, then lay the leaves on an open cotton towel. Wrap it all up by pulling the corners together crosswise and the lettuce will stay crisp for a week. Inexpensive cotton bags are often sold at natural food stores, or order by mail from ***Clothcrafters.***

♦ CELLOPHANE BAGS. Made from natural cellulose, these are a biodegradable alternative to plastic bags. Not as sturdy, but easier to assimilate back into the environment. Natural food stores usually carry them, or order from ***Earth Care Catalog*** or ***Seventh Generation.***

⬇ PLASTIC BAGS. Don't use or wash and reuse—they are made from nonrenewable resources and not biodegradable.

Fragrances

⬆ MAKE YOUR OWN. Buy essential oils from your natural food store and add a few drops to one ounce of vegetable oil or vodka to dilute. All natural essential oils are made from plants that have not been sprayed with pesticides, as the pesticides would concentrate in the oil during the distilling process.

⬆ ORGANICALLY GROWN INGREDIENTS. **Alexandra Avery** and **Nature's Colors** have perfume blends made with essential oils and other organic ingredients. Your natural food store will carry these and other brands.

⬇ HARMFUL INGREDIENTS. Most perfumes and colognes (even expensive ones, and particularly the cheap ones) are made with artificial fragrances and other petrochemical ingredients. Allergic reactions to these fragrances are commonplace.

See also BEAUTY AND HYGIENE.

Furniture

⬆ USED. Most sustainable would be to refurbish already existing furniture by recovering it with natural fiber upholstery, or sanding it down and giving it a new natural finish. Garage sales, storage auctions, and flea markets are great sources for inexpensive buys. My desk is a solid mahogany library table I got for thirty dollars at a garage sale, my oak desk chair came from a salvage yard for twenty-five dollars, our solid oak dining table was fifteen dollars at a garage sale, our cotton-stuffed sofa (which needs recovering) was fifty dollars at a storage auction. Antique and used furniture dealers have a lot of interesting and attractive items for your home.

⬆ SUSTAINABLY HARVESTED WOOD. SCS certified wood from the Menominee Indian Reservation is used in **The Gehry Collection** office furniture by The Knoll Group.

↑ SOLID WOOD. Solid-wood furniture can be found at many unfinished furniture stores, and can be protected with a natural finish (see PAINT). Wood furniture should be checked carefully to make sure it is indeed solid wood. Often the front will be wood and the backs, sides, inside shelves, and drawer bottoms will be particleboard or plywood (see below). ***Heart of Vermont*** has a good selection of simply styled bed frames and other furnishings made by local craftspeople from local hardwoods and nontoxic glues.

↑ NATURAL FIBERS. Upholstered furniture should be covered with a natural fiber fabric that does not have any kind of finish. Most upholstery fabrics have a stain-resistant finish, but these are clearly marked, so you should be able to find some without the finish. On my recent search for fabric to recover my sofa, I found so many fabrics without finishes that I stopped making notes about them—I had about one hundred choices at a single store. Find an upholsterer who will fill the cushions with cotton or wool batting or feathers.

↑ RECYCLED SYNTHETIC FIBERS. Upholstered furniture stuffed with spun polyester fill made from recycled PET plastic soda bottles.

↓ SYNTHETIC FABRICS. Made from nonrenewable resources, not biodegradable, and usually have formaldehyde-based finishes.

↓ PARTICLEBOARD AND PLYWOOD. These wood products are made from chips and sheets of wood, held together by resins. Particleboard is notorious for its urea-formaldehyde resin which offgasses from the large amounts of formaldehyde used (plywood uses less volatile resins). If you have to have particleboard, use a vapor barrier sealant from **AFM.**

For a listing of furniture and other home furnishing products, subscribe to *Interior Concerns* newsletter and get a copy of their *Interior Concerns Resource Guide* which lists products, manufacturers, consultants, and other resources.

Furniture and Floor Polish

↑ MAKE YOUR OWN. Use a soft cloth to apply mayonnaise or a mixture of olive oil and lemon juice or vinegar. Polish until absorbed.

↑ NATURAL/RENEWABLE OR HYBRID-NATURAL INGREDIENTS. May be available in some greenstores and green catalogs. **Livos** polishes and waxes (imported from Germany) can be ordered by mail from ***Natural Choice.***

↑ NONTOXIC/NONRENEWABLE INGREDIENTS. May be available in some greenstores and green catalogs. **AFM** makes nontoxic polishes and waxes.

↓ TOXIC/NONRENEWABLE INGREDIENTS. Don't use products that contain aerosol propellants, ammonia, detergents, synthetic lemon or other fragrance, nitrobenzene, phenol, or plastics. Note "DANGER" warning on package label.

See also CLEANING PRODUCTS.

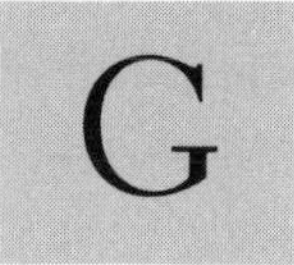

Garden

↑ ORGANIC GARDENING. Gardening in harmony with nature takes more awareness, study, and planning than gardening with artificial fertilizers and toxic pesticides. It's a skill to hone over a lifetime—there is much to learn. But there are also plenty of resources with information, tools, and supplies. Order these catalogs and dig in: ***Ecology Action, Gardener's Supply, Gardens Alive!, Harmony Farm Supply, The Kinsman Company, The Natural Gardening Company, Necessary Trading Company, Peaceful Valley Farm Supply,*** and ***Smith & Hawken.*** While these catalogs have good organic gardening books, ***Acres, U.S.A.*** and ***agAccess*** specialize in books on all aspects of organic gardening and agriculture.

↑ ORGANICALLY GROWN OPEN-POLLINATED SEEDS. Seeds are the foundation of our food supply and the source of most plant life on earth. Over billions of years, nature evolved plant types that were ideally suited to the various climates and soil conditions around the world. In the last ten thousand years, farmers have been selecting those varieties that produce the hardiest plants and best foods for local growing conditions, and carefully saving the finest specimens for seed stock. Throughout this natural selection, all the plants were naturally open-pollinated and each was slightly different, having its own unique combination of genes.

At the turn of this century, new biological discoveries allowed scientists to make deliberate crosses between two varieties of the same species. Instead of allowing the plants to cross-pollinate with each other as they would in a garden, hybridizers hand-pollinate each flower with pollen from the same plant. This inbreeding is repeated for as many as ten or twelve generations until they are genetically uniform. When seeds from the inbred line are planted, they produce the identical plants expected in our consumer world, but they are often sterile and never display the productivity of their parents.

Ninety-nine percent of the seeds used today by agribusiness, organic farmers, and home gardeners are hybrids that produce plants that cannot sustain themselves without human intervention. For sustainability, we need to grow plants with their natural ability to reproduce, and collect and distribute their seed far and wide. Order these catalogs to purchase organically grown open-pollinated seed: ***Abundant Life Seed Foundation, Corns, Deep Diversity, Ecology Action, Garden City Seeds, Goodwin Creek Gardens, Halcyon Gardens, J. L. Hudson, Seedsman, Native Seeds/SEARCH, Seeds of Change, Talavaya Seeds,*** and ***The Tomato Seed Company.***

🡅 SUSTAINABLY HARVESTED WOOD. Some wood garden furniture is now being made with domestic and rain-forest woods that are sustainably harvested. ***Smith & Hawken*** has a good selection.

Glass Cleaners

🡅 MAKE YOUR OWN. My favorite glass cleaner is half vinegar and half water, applied with a soft cloth or pump-spray bottle. This works so well that some big corporations are selling it in the supermarket in a plastic bottle with a little green dye added.

🡅 ORGANICALLY GROWN NATURAL/RENEWABLE INGREDIENTS. Not yet available to my knowledge, but you can use VINEGAR made from organically grown ingredients in your homemade formula.

🡅 NATURAL/RENEWABLE OR HYBRID-NATURAL INGREDIENTS. A few are available in natural food stores, greenstores, and green catalogs.

🡇 TOXIC/NONRENEWABLE INGREDIENTS. Don't use products that contain ammonia, detergents, artificial dyes, or aerosol propellants even though there are no warnings required on product labels.

See also CLEANING PRODUCTS.

Grains

🡅 ORGANICALLY GROWN. Common and unusual grains are easy to find in natural food stores and catalogs. To take an extra step into sustainability, look for grains that are open-pollinated, which means they have been grown from seed that can reproduce, rather than from hybrid seed. Open-pollinated seed can be collected from year to year and sustain an ongoing crop, whereas hybrid seed must be created anew and purchased from seed companies. Look for open-pollinated grains at your natural food store or order from ***Deer Valley Farm*** or ***Gold Mine Natural Food Company.*** Some whole grains are also biodynamically grown.

⬆ NATURAL. Whole grains can be purchased in bulk in any natural food store or catalog.

See also FOOD.

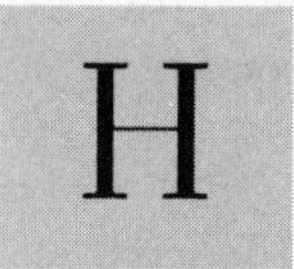

Hairbrushes and Combs

⬆ WOOD AND NATURAL BRISTLES. Wooden brushes with natural bristles are commonly available wherever brushes are sold. Wood combs are often sold in natural food stores or you can order them by mail from ***Traditional Products Company*** or ***Heart of Vermont.***

⬇ PLASTIC. Plastic brushes and combs are made from nonrenewable resources and are not biodegradable.

Hairspray

⬆ GET A GOOD HAIRCUT. A sustainable style doesn't require hairspray.

⬆ NATURAL/RENEWABLE AND HYBRID-NATURAL INGREDIENTS. Your natural food store should have a selection of natural hairsprays in pump-spray bottles.

⬇ TOXIC/NONRENEWABLE INGREDIENTS. Regular hairspray is made entirely of nonrenewable ingredients.

See also BEAUTY AND HYGIENE.

Heating Systems

🡅 SOLAR. The most sustainable heat for your home is warmth from the sun. For centuries homes have been designed to take advantage of passive solar energy. In ancient Greece entire towns were designed so that all the homes could have optimal southern exposure to be cool in summer and warm in winter. Contact the energy organizations listed in the Directory of Organizations for more information.

🡅 ELECTRIC SPACE HEATERS. Rather than heating your whole house, at times all you may need is a spot of warmth in the morning in a cold bathroom, or some heat for your feet under your desk. For these times, have a small space heater on hand (sold in most hardware stores). Heaters with ceramic elements are particularly efficient.

🡅 WOOD AND PELLET STOVES. A wood stove can heat your home very well if you live in an area where wood is affordable, abundant, and can be obtained in a sustainable way. Unlike fossil fuels, wood burns clean when completely combusted, returning the same amount of carbon dioxide to the air as would have been released if the tree decayed on the forest floor.

In the past inefficient wood stoves created quite a bit of air pollution. After Americans turned in droves to inexpensive wood heat in the energy-expensive mid-1970s, the concentration of wood smoke pollution became a major problem in some areas. In 1986 the EPA joined with environmentalists, scientists, state agencies, and representatives from the wood heating industry to forge new regulations for air emissions and energy efficiency. EPA regulations now prohibit the manufacture and sale of wood stoves that cannot pass a stringent particulate emissions test, done in one of eight EPA-certified testing laboratories. Beginning in 1988 emission limits were phased in and grace periods were allowed during which noncomplying stoves could be sold. As of July 1, 1992, all stoves sold must meet emissions requirements and be EPA certified. A label attached to each stove states "Meets EPA particulate matter (smoke) control requirements . . ." and shows the amount of emissions and energy efficiency in a graphic illustration similar to the EnergyGuides.

Modern wood-burning stoves rely on three different technologies for their performance. Traditional wood stoves have either been redesigned to be more airtight or smoke-burning catalytic converters have been added. Emissions from noncatalytic stoves manufactured after July 1, 1990, may not exceed 7.5 g/hour. Emissions from catalytic stoves manufactured after July 1, 1990, may not exceed 4.1 g/hour. This is an 85 percent reduction of emissions over models sold in 1986.

The most significant new development is the pellet stove that burns small cylinders made from recycled wood chips, sawdust, bark, and other wood scraps.

Because the pellets contain only about 3 percent moisture (seasoned cordwood contains about 25 percent), they have the ability to burn very cleanly in their specially designed stoves.

Pellet stoves are a lot easier to install, start, and feed than a traditional wood stove. A hopper near the top of the stove holds about 80 pounds of pellets. They slowly trickle down into a small firebox where an electric fan forces air into the fire to make a small, hot blaze. Another electric fan carries the heat into the room. They burn so completely that often the only flue needed is a small class-L vent (similar to the kind used for natural gas furnaces). Two models emit only 0.5 g/hour particulates. Electricity use is no more than the continuous burning of two 100-watt light bulbs.

Pellet stoves may be the answer for burning regionally appropriate fuels instead of fossil fuels or wood. Stovemakers are already adapting their stoves to burn pellets made from corncobs from Nebraska, walnut shells from Missouri, and sunflower hulls from Minnesota.

♦ ELECTRIC HEAT PUMPS. For heating a whole house, a heat pump can deliver up to three times more heat energy than is contained in the electricity used to power it. Heat pumps use electricity to move heat from one area (called the source) and to another area (called the sink). A CFC or HCFC refrigerant with a boiling point far below room or outdoor temperatures is contained in a sealed coil and exposed to a heat source. The refrigerant absorbs heat from the source and boils as it warms up, turning into a gas. As a gas, the refrigerant is then pumped through a compressor and into another sealed coil exposed to the sink. As the refrigerant condenses, it releases heat which is absorbed by the sink. Most common sources of sinks are outdoor air providing heat for indoor air, though heat pumps can also be used with water. A heat pump operated in reverse acts as an air conditioner. Central heat pump systems can be used year-round for heating and cooling.

The efficiency of heat pumps is expressed by the ratio of the heat delivered to the sink to the heat content of the electricity consumed to power the motor. This ratio is called the Coefficient of Performance (COP). The efficiency of the unit is affected by the temperature of the source, so if it's very cold outside, the heat pump will be less efficient. They are most appropriate in warmer climates where the temperature seldom goes below 30 degrees in the winter and summer cooling is needed. The most efficient heat pumps are listed in *Consumer Guide to Home Energy Savings*.

♦ ENERGY-EFFICIENT GAS AND OIL HEATING SYSTEMS. If we are to burn nonrenewable fossil fuels to produce heat, it should be in a way that makes most efficient use of the fuel.

The efficiency of a gas or oil heating system is a measure of how efficiently it converts fuel into useful heat. There are two types of efficiency. Combustion efficiency tells you how efficient the system is while running (like the miles per gallon your car gets while cruising at fifty-five miles per hour on the highway). But a more accurate estimation of the actual fuel use is the annual fuel utilization efficiency (AFUE). This is a measure of the system's efficiency that accounts for start-up and cool-down and other operating losses that occur in real-life conditions. AFUE is like your car mileage in around-town stop-and-go traffic.

When shopping for a gas or oil heating system, look for the AFUE on the EnergyGuide label to determine energy efficiency. The most efficient gas furnaces available have an AFUE of more than 96 percent, the best gas boilers achieve efficiencies up to about 90 percent, and oil-fired hot-water boilers are available with efficiencies up to 91 percent. The most efficient brands and models are listed in *Consumer Guide to Home Energy Savings.*

If you have central heating, a programmable clock thermostat can also help you save energy by keeping the heat lower at night and during the daytime hours when you are away. This can save 20 to 40 percent of energy during the day and 10 to 14 percent at night. One disadvantage to using gas is combustion by-products, particularly carbon monoxide. In California by law we have received warnings that natural gas contains benzene, which is known to cause cancer. When using a gas appliance of any kind it's important to watch for gas leaks and to adequately vent combustion by-products.

↓ ELECTRIC HEATING SYSTEMS. These directly convert electric current into heat. They are highly inefficient and expensive to run, and therefore are not recommended.

↓ FIREPLACES are not recommended for home heating—80 to 100 percent of the heat generated goes up the chimney.

Herbs and Spices

↑ GROW YOUR OWN. Even if you don't have time or space for a garden, you can generally grow herbs on a windowsill. It's nice to grow something you can eat—even if it's just a sprinkle in your salad—and the fresh flavor is wonderful.

↑ ORGANICALLY GROWN. Natural food stores and catalogs usually carry a good variety of dried whole herbs and spices in bulk. Grind them yourself at home. And look for fresh organically grown herbs at your natural food store and farmers' market. Dried biodynamic culinary herb mixtures can be ordered by mail from ***Meadowbrook Herb Garden.***

⬆ NATURAL. There are many additive-free herb and spice seasoning mixes in natural food stores, import stores, gourmet emporiums, and supermarkets. Check the labels carefully to make sure all the ingredients are recognizable foods, and no anticaking agents have been added.

⬇ IRRADIATED. Do not buy irradiated herbs and spices.

See also FOOD.

Insect Repellents

⬆ WEAR PROTECTIVE CLOTHING. Putting on a loose-fitting shirt on summer evenings will keep some insects off your skin. If you're going outdoors into a heavily infested area, look for insect-protective clothing in an outdoor outfitting store or catalog.

⬆ PUT UP SCREENS OR NETTING. Put mosquito netting around your bed, or install screens on windows and doors.

⬆ CITRONELLA CANDLES. These can be purchased anywhere housewares are sold.

⬆ NATURAL/RENEWABLE REPELLENTS. You can rub various different natural substances on your skin to repel insects. The quickest and easiest repellent is vinegar—just splash a little on exposed skin or dab it on with a cotton ball (this really works; it's all I ever use). Or you can use essential oil of citronella or pennyroyal (available at natural food stores), diluted with vodka or vegetable oil (a few drops to one ounce of either). Your natural food store should have some herbal repellents; one good one you can order by mail is ***Naturally Free.***

⬇ TOXIC/NONRENEWABLE. Don't use insect repellents that contain DEET. Reported negative health effects include brain disorders, slurred speech, difficulty walking, and tremors. Some children have even died. The warning label cautions that it "may damage furniture finishes, plastics, leather, watch crystals, and painted or varnished surfaces including automobiles." It doesn't seem prudent to me to rub something on your skin that could take the paint off your car.

Insulation

The insulating ability of insulation is indicated by the "R-value," which stands for resistance to heat flow. R-1 is roughly equivalent to a single pane of glass. The higher the R-value, the greater the insulating properties. In a temperate area the recommended R-values are R-50 in ceilings, R-32 in walls, R-24 in floors above unheated areas, and R-16 in foundations.

🡅 NONTOXIC MINERAL-BASED INSULATION. An insulation often used by natural builders is **Air-Crete Foamed-in-Place Insulation.** It is a cementitious magnesium-oxide foam that hardens after application to an inert hard sponge.

🡅 BLOWN-IN CELLULOSE INSULATION FOR WALLS. Made from recycled newspapers and sodium borate, blown-in cellulose is cheaper than fiberglass, and has greater insulating properties.

🡅 FIBERGLASS INSULATION MADE WITH RECYCLED GLASS. **Manville Building Insulation** is certified by Scientific Certification Systems to contain a minimum of 15 percent post-consumer remelted glass and 5 percent pre-consumer glass (and it may contain up to 30 percent recycled glass).

🡅 BLOWN-IN FIBERGLASS AND FIBERGLASS BATTS. Frequently used for insulation and generally considered safe for all but the most sensitive people. Batts won't, however, perform to the rated R-value unless they are correctly installed to fill in all the nooks and crannies—fiberglass can be blown in to ceilings.

🡇 UREA-FORMALDEHYDE FOAM INSULATION (UFFI). The Consumer Product Safety Commission (CPSC) banned the use of UFFI in residences and schools in 1982, but one year later this ban was overturned by the United States Court of Appeals. The CPSC feels the court's decision was based on legal and factual errors and continues to warn consumers that available evidence indicates there are risks associated with this product.

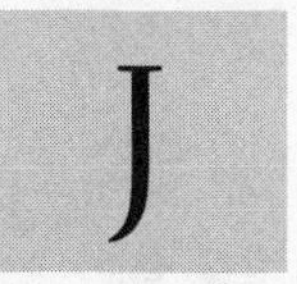

Jams, Jellies, Preserves, and Fruit Spreads

🡅 ORGANICALLY GROWN. Easy to find in natural food stores and catalogs, naturally sweetened with fruit juice, syrup honey, or other organic sweeteners.

🡅 NATURAL. All jams, jellies, and preserves are additive-free, but most contain sugar. A popular alternative is fruit spreads, which taste virtually identical but are sweetened with concentrated fruit juice. Look for fruit spreads practically everywhere jams and jellies are sold.

See also FOOD.

Juice

⬆ MAKE YOUR OWN. Even though bottled juice may be made with organic ingredients, the heat from the bottling process destroys so much of the vitality and nutrition in the juice, that it ends up being practically flavored water in comparison to fresh. Get a good juicer and make your own fresh juice.

⬆ ORGANICALLY GROWN. There are many refreshing juice flavors now made from organically grown fruits. They're easy to find in natural food stores and catalogs.

⬆ NATURAL. Many natural juices are available at your supermarket and natural food store. Read the ingredients to make sure you are buying 100 percent juice and not a "juice-drink" or "cocktail" with added sugar.

See also FOOD.

Kitchen Linens

⬆ NATURAL FIBERS. All kitchen linens—towels, pot holders, aprons, etc.—are available in natural fibers at cookware shops and import stores.

See also TEXTILES.

Laundry Soap and Detergent

⬆ CHOOSE NATURAL FIBERS. Detergents were developed especially to clean synthetic fibers, and are unnecessary for natural fibers such as cotton, linen, silk, and wool, which can be washed using soap.

↑ RINSE WITHOUT DETERGENT OR SOAP. You don't always need to use soap or detergent to get clothes clean. If you need to wash clothes to freshen them or remove perspiration or odors, and not remove dirt, a cup of plain baking soda or vinegar per washload will do the trick.

↑ USE SOAP. Use a plain powder or liquid, or grate bar soap. One problem with soap is that it leaves a residual scum on fabrics when used in hard water. This can be eliminated by using a WATER SOFTENER. A very effective soap-based laundry compound can be ordered from ***Cal Ben Soap Company.***

↑ NATURAL/RENEWABLE OR HYBRID-NATURAL INGREDIENTS. Many brands are available in natural food stores, greenstores, and green catalogs.

↑ NONTOXIC/NONRENEWABLE INGREDIENTS. Many brands are available in natural food stores, greenstores, green catalogs, and hardware stores.

↓ TOXIC/NONRENEWABLE INGREDIENTS. Don't use products that contain detergents, fluorescent brighteners, or artificial fragrances. Warnings on package labels range from "CAUTION" to "DANGER"—detergents cause more household poisonings than any other household product.

See also BLEACH and CLEANING PRODUCTS.

Light Bulbs

The energy efficiency of a light bulb is a measure of how much energy is required for the creation of the light output of the bulb. Obviously, the less energy required to create a particular amount of light, the more efficient the bulb. To determine the efficiency, the total light output of the lamp (measured in lumens) is divided by the full wattage of the lamp and the ballast. This gives a lumens per watt efficiency, with a higher number being more efficient.

If you use a dimmer switch, you can save energy even with your incandescents, but at the expense of the amount of light you receive. In most cases you'll save more energy and have more light with an energy-efficient light bulb.

↑ COMPACT FLUORESCENT BULBS. Compact fluorescents are the most energy-efficient light bulbs available, converting to light about 40 percent of the energy used. If we all used compact fluorescents in our homes and businesses, we could eliminate the need for all the nuclear power plants operating in the United States. They also last longer, with life spans ranging from 7,500 to 100,000 hours instead of the 750 to 1,000 hours for a standard incandescent bulb. Although these bulbs cost ten to twenty-eight dollars, they are about half as expensive over their lifetimes as incandescent bulbs because of the greatly reduced energy use. An 18-watt compact fluorescent gives the same light as a 75-watt incandescent bulb and lasts thirteen times longer.

While the energy savings are great, compact fluorescents have some real limitations. They will not work with dimmers or with some timers, motion sensors, and photosensors, and they might be too big to fit your fixture. You can put a compact fluorescent in a three-way switch, but it will only turn on or off.

More importantly, according to Scientific Certification Systems, tests done at Lawrence Berkeley Laboratories and GTE/Sylvania show that the claims made by compact fluorescents are likely to be misleading when the bulbs are used in the real world. Claims of efficiency, bulb life, and brightness of light are determined in a laboratory where optimum conditions are used. The claims would be true if your home had the same optimum conditions, but rarely is this the case.

The efficiency of the bulb (losses up to 20 percent) can be affected by the direction in which it points and the ambient temperature around the bulb. Mercury is needed to create fluorescent light, which is in the bulb in the form of a drop of solid mercury and mercury vapor. When the bulb has the base down (tube pointing up) the mercury rolls down near the ballast and vaporizes, making the bulb overburn. If the lamp gets too cold, the mercury vapor solidifies and the lamp underburns. For greatest efficiency, the ideal orientation for a compact fluorescent is base up (with the tube pointing straight down or at an angle) and the ideal temperature is 77 degrees.

These tests also show that the life of compact fluorescent bulbs likely will not be as long as claimed. Standard industry testing protocol for fluorescent bulbs is to determine life by a test of three hours on and twenty minutes off. Turning the light on and off more frequently wears down the filament faster and the bulb will not last as long. Compact fluorescents are most efficient where they will stay on for an hour or more, so in areas where you flick the light on and off (like the bathroom), a compact fluorescent might not be the most appropriate choice.

And you may not get as much light as you expect from a compact fluorescent. Package labels indicate a lumens equivalent, but 75 watts from an incandescent is not the same as a 75-watt-equivalent fluorescent. The difference has to do with the type of light emitted. Incandescents emit a point source light, where all the light comes from a single point. Fluorescents have a diffused source because the light is coming from all different points along the tube. Therefore the amount of light coming from the nose of the tube, for example, is less than the light coming off the side of the tube. So even though you have the right wattage, you might not get all the light you expect. In addition, over the life of the bulb, compact fluorescents will lose lumen output by about 20 percent as it reaches the end of its rated life.

These limitations don't mean you shouldn't use compact fluorescents. The ideal application for these bulbs are in places where they are on all day or night (such as the hallway or laundry room of an apartment building, or an office) and the temperature is moderate and comfortable.

Compact fluorescents come in two types: integral and modular, the distinction having to do with the ballast. In order to be operable, all fluorescent lights need a ballast to jolt the electric current and get the gas inside glowing. Ballasts are either magnetic or electronic (electronic ballasts are quieter, turn on without flickering, and are more expensive).

If you want an electronic ballast, you need an integral model (they also come with magnetic ballasts) where the ballast and tube are built into a single unit. Modular compact fluorescent lights have separate ballasts (magnetic only) and tubes. Modular ballasts (sometimes called adapters) usually screw into standard light bulb sockets, or you can also buy light fixtures that are hard-wired with a ballast built into the fixture. The advantage to modular units is that you don't have to replace the ballast every time the tube burns out. While tubes last about ten thousand hours, ballasts may last fifty thousand hours or longer, and replacement lamps cost much less than buying a whole new integral unit.

One objection many people have to using fluorescents is that the color of the light is too blue. But that has improved—you now have your choice of color temperature (given in degrees Kelvin to indicate the color appearance). Incandescent lamps are about 2,700 degrees K and daylight is about 5,500 degrees K. Compact fluorescents come in 2,700 degrees K for incandescentlike light, 3,500 degrees K for white light, and 4,100 degrees for cool-white light. The CRI (color rendering index) rating tells how much the color of an object will be distorted under the light (perfect rating—no change—is 100), and can be used to compare bulbs with the same color temperature. A typical incandescent has a CRI of about 92 to 100 and a standard "blue" fluorescent tube has a CRI of about 52. A warm white compact fluorescent has a CRI of about 82 to 85.

Compact fluorescents come in extremely small wattages—9 to 26 watts instead of the 40 to 100 watts we are accustomed to buying. It will say on the package how many incandescent watts the bulb is equivalent to, so just buy the same watts-equivalent you usually buy.

Some compact fluorescents use small quantities of radioactive material in them, so make sure the ones you buy do not. The U.S. Nuclear Regulatory Commission requires that the bulbs that do contain radioactive materials be so labeled, but this information may be in very fine print, so look carefully. Those that have an electronic ballast are nuclear-free.

Green Seal has a certification standard for compact fluorescent lamps that requires warm incandescentlike light (CRI of no less than 80) and rapid startup. Efficiency required ranges from 40 to 62 lumens per watt (depending on wattage) and the average minimum rated product life must be 8,000 hours.

Green Seal also requires companies to provide information on the label, including a wattage rating that compares the compact fluorescent to an incandescent's light output, explanation of proper and improper installations, and the energy and cost savings over the life of the bulb. More than twenty power companies in the United States have adopted Green Seal's standard for compact fluorescent lights, stipulating that manufacturers must meet these standards in order for their products to qualify for subsidies, such as consumer rebates. Compact fluorescents made by **General Electric** and **Lights of America** have been certified by Green Seal.

Compact fluorescent bulbs were a specialty catalog item only a few years ago. Now they are sold in all my local hardware stores. If you have trouble finding them, you can order by mail from ***Real Goods*** or ***Seventh Generation.***

⬆ HALOGEN BULBS. A more practical choice of energy-saving bulbs for household use are halogens. If you don't have a light that needs to burn for hours, halogens coupled with motion sensors (to turn the lights off when you forget) might be the combination that saves you the most energy.

Like compact fluorescents, halogen bulbs use less energy and last longer, although the savings are not as great. About 15 percent of the energy used is converted to light (giving about 25 percent more light than an incandescent) and they last about two thousand hours. Because halogens have a bright white light, lower wattages can often be used, and where light is needed in a precise area, halogens can be even more efficient than fluorescents because they can be more accurately focused. And because they are so small, fewer resources are needed to manufacture the bulbs and the fixtures.

According to the American Lighting Institute, the light from a halogen bulb most resembles sunlight. Because of its excellent color-rendering properties it is a good choice for the kitchen and for grooming. Everything looks more vibrant and sparkling under halogen light.

Halogens are most commonly used in spot and track lights, although **Sylvania** makes a great halogen replacement for incandescents that is about the same size and shape (though noticeably heavier). Halogen bulbs are sold in many hardware stores, and particularly in specialty lighting shops as they are often the choice of decorators.

A word of warning, however. Unlike other types of light bulbs, halogens—like sunlight—emit ultraviolet radiation. Recent research from Australia, Italy, and Great Britain suggests that halogen bulbs can cause sunburn and, over a long period of time, increase the risk of skin cancer. I, for one, am not going to remove all the halogen bulbs from my home, but you might not want to sit directly under one eight hours a day. The FDA has requested that halogen lamp manufacturers provide glass shields for their lamps—ask your local lighting dealer for shields.

↑ ENERGY-SAVING INCANDESCENT BULBS. These are frequently sold in supermarkets and have a very small percentage of savings. They are better than an incandescent, but not as efficient as a halogen or compact fluorescent.

↑ LONG-LIFE BULBS. These do not save energy during use, but they have greatly extended life spans, and will in the long run save on manufacturing resources. One long-life bulb (lasts three times longer than an ordinary incandescent), **EcoWorks Ecological Lightbulb,** also uses 10 percent less energy and is made by a nuclear-free company.

↑ FULL-SPECTRUM BULBS. These do not save energy, but have a color range that more closely duplicates sunlight (though not exactly).

↓ INCANDESCENT BULBS. Not very efficient. Don't use unless there is no other option.

M

Meat and Poultry

↑ EAT LESS MEAT. As a nation we eat much more protein than we really need for good health. There are logical arguments pro and con strict vegetarianism, but everyone and the environment can benefit from eating less meat and more grains, legumes, nuts, vegetables, and fruits.

↑ ORGANICALLY GROWN AND HUMANELY RAISED. Many natural food stores now carry fresh or frozen meat and poultry from sources whose animals have been pasture grazed and raised without hormones or other chemicals to stimulate or regulate growth or tenderness, and without drugs or antibiotics. And look for wild game that has been hunted (in sustainable numbers, of course) in its wild habitat.

↓ FACTORY FARMS. Most meat, poultry, and other animal products on the market today come from factory farms that raise animals as "biomachines." The animals are confined in dark, crowded quarters, fed a diet high in drugs and chemicals and low in nutrients, and they are deprived completely of exercise and fresh air. The results are highly stressed, deformed, and diseased animals, which hardly qualify as high-quality foods. And they are not even animals as nature made them; thanks to the modern science of genetic manipulation, we are sold animals that grow faster (for efficient, high-yield production), but are of poorer quality.

See also FOOD.

Milk

⬆ ORGANIC. This is becoming more available from local dairies and at natural food stores.

⬆ ORGANICALLY GROWN SOY MILK. This milk substitute is commonly sold in natural food stores and catalogs (an environmental drawback is that it generally comes in an aseptic package that ends up in a landfill).

⬆ RAW. Raw milk is the unprocessed liquid that comes directly from a cow or goat. It is graded according to levels of bacteria found in it. Grade A has the lowest bacteria count and in addition cannot, by law, contain detectable antibiotic residues. "Certified milk—raw" means that the milk conforms to the latest requirements of the American Association of Medical Milk Commissions—it does not mean that it is certified organically produced.

⬆ GLASS BOTTLES. If possible, buy milk in glass bottles. Some local dairies will deliver it to your door. Glass bottles are returnable, refillable, and recyclable, unlike paper cartons (which might also leach minute amounts of toxic dioxin into the milk) or plastic bottles. Sometimes you can buy glass-bottled milk that has not been fortified or homogenized (see below).

⬇ PASTEURIZED, HOMOGENIZED, FORTIFIED. About 98 percent of all the milk we drink comes from factory farms, and contains residues of pesticides, antibiotics, hormones, or tranquilizers that have been fed to the cows, and antiseptic solutions that have been used to clean cow udders before milking. Fortification with vitamin A or D adds propylene glycol and BHT. In addition, the natural structure of the milk is changed when it is pasteurized (heated to destroy bacteria) and homogenized (mixed under pressure to reduce fat particles to a uniform size and texture).

See also FOOD.

Mold Cleaners

⬆ KEEP ROOMS DRY AND LIGHT. Mold is a living organism that will only grow in dark, damp places, so if you have a recurrent mold problem, bring more light, heat, or fans to move the air.

↑ LET TEXTILES DRY BEFORE STORING. Hang up wet towels after bathing to let them dry before throwing them in the hamper. Hang clothes so there is space between them, and if you don't launder clothing that is damp with perspiration, at least allow it to dry before putting it back in the closet.

↑ USE HEAT. For major mold problems, put a portable electric heater in the room, and turn it to the highest setting. Close the door and let it bake all day or overnight. The mold will dry up into a powder that brushes right off. For concentrated areas, use a hand-held dryer to dry the mold in just a few minutes. This is not the most energy-efficient method, but it will definitely solve the immediate problem.

↑ MAKE YOUR OWN. Sprinkle baking soda on a damp sponge and wipe a moldy surface. Mold will come off fairly easily. Don't rinse—a residue of baking soda will inhibit mold growth (works best on light-colored surfaces).

↑ NATURAL/RENEWABLE, HYBRID-NATURAL, OR NONTOXIC/NONRENEWABLE INGREDIENTS. These are not generally available, but there are a couple of nontoxic products by ***AFM***. Their **Safety Clean** will remove existing mildew, and their **X158 Mildew Control** provides mildew resistance and preservation against microbial attack.

↓ TOXIC/NONRENEWABLE INGREDIENTS. Don't use products that contain formaldehyde, phenol, pentachlorophenol, or kerosene. Note "DANGER" warning on package label.

See also CLEANING PRODUCTS.

Mothballs

The moths that eat woolens are a specific variety of clothes moth that is too small to see. It is the larvae of these moths that eat fiber.

↑ REMOVE MOTH EGGS. Kill moth eggs before they hatch by placing items in the sun, by running them through a warm clothes dryer, or by washing. This should be done when the item is first purchased and at periodic intervals while being stored.

↑ PROTECT. Store uninfested items in clean condition in airtight containers, such as paper packages or cardboard boxes, with all edges carefully sealed with paper tape.

🡅 NATURAL/RENEWABLE REPELLENTS. Herbal repellents are very effective and smell nice, too. Lavender, rosemary, mint, and peppercorns all repel moths, but the classic repellent is cedar. Cedar can be purchased in many forms—oil, chips, blocks, balls, and drawer liners—in almost every place mothballs are sold. ***Clear Light*** makes cedar sachets and essential oils from cedar branches trimmed from trees in such a way that enhances the growth of the trees (these are available in many gift shops or can be ordered by mail), and ***The Hummer Nature Works*** makes cedarwood items from deadwood timber left lying around the countryside or trees that have fallen because of weather conditions or disease.

🡇 TOXIC/NONRENEWABLE. Don't use mothballs made of paradichlorobenzene—the warning label cautions against "prolonged breathing of vapor," but how can you use mothballs without breathing their vapor? The odor of mothballs hidden in a closet can permeate your entire home and increase to high levels if there is not adequate ventilation. The vapors from mothballs are also absorbed by clothing and blankets, and can be very strong when you use these items.

Motor Oil

🡅 RE-REFINED. Green Seal has a certification standard for re-refined motor oil. It requires that 25 percent of the oil base stock (the oil itself without the performance additives) be re-refined. As of January 1, 1995, the re-refined content will be increased to 40 percent. Containers must be made of 10 percent post-consumer materials as of January 1, 1994, and 25 percent as of January 1, 1996. Ask for re-refined oil at your local auto supply dealer.

Mouthwash

🡅 PREVENT BAD BREATH. Brush and floss regularly and you won't need mouthwash to freshen your breath.

🡅 ORGANICALLY GROWN INGREDIENTS. Your natural food store should carry **Weleda Mouthwash** (or you can order from ***Meadowbrook Herb Garden***), which contains biodynamically grown herbs.

🡅 NATURAL/RENEWABLE AND HYBRID-NATURAL INGREDIENTS. You'll find many natural brands at your natural food store.

See also BEAUTY AND HYGIENE.

Nutritional Supplements

🡅 EAT FRESH, WHOLE, UNPROCESSED, ORGANICALLY GROWN OR BIODYNAMICALLY GROWN FOOD, GROWN IN FERTILE, WELL-MINERALIZED SOIL. In a completely sustainable world there would be no need for nutritional supplements. And if you have a good source of food (homegrown or locally grown) that has been grown in good soil, you probably don't need nutritional supplements. But in the reality of today's world, almost everyone needs some kind of supplement at some time to augment what is missing in diet, to satisfy a deficiency, or support physical stress.

🡅 WHOLE FOOD. The supplements that come closest to obtaining nutrients directly from foods are supplements that are foods. They come in three forms: highly nutritious, concentrated foods (such as bee pollen); powdered concentrates of foods and herbs with the moisture removed (as in barley-juice powder); and an isolated part of a food (wheat germ, for example). Herbs are also increasingly being used as a source of minerals, vitamins, and other nutrients. Your natural food store should have a good selection of these supplements, as well as information on how to choose those that are right for your needs.

🡅 NATURAL. Many supplements call themselves "natural" because they come from renewable resources and are additive-free; however, at best they are hybrid-natural. Most natural supplements on the market today are either fortified (a very small amount of low-potency natural vitamins mixed with high-potency synthetic vitamins) or are synthetic vitamins in a very small amount of natural base (the label will say something like "in a natural base containing . . ."). All supplements sold at natural food stores are at least additive-free, even though many are hybrid-natural.

🡇 SYNTHETIC. The heavily advertised national brand nutritional supplements are for the most part synthetic vitamins with artificial additives. Even many natural supplements, though additive-free, are made with synthetic nutrients derived from petrochemicals. There are only twelve pharmaceutical manufacturers in the world (five in the United States) that make vitamins; all the few thousand vitamin brands buy from them. Nutritional supplements are a consumer industry that operates hand-in-hand with the nutritionally deficient processed food industry.

Even though natural and synthetic supplements are chemically identical, natural supplements do have molecular, biological, and electromagnetic differences that produce greater levels of biological activity; they are better utilized by our bodies than synthetic forms. This quote from the book *The Body Electric* by Robert O. Becker, M.D., explains a crucial difference: "All organic compounds . . . are identified by the way they bend light in solution. The dextrorotatory (D) forms rotate it to the right while levorotatory (L) isomers refract it to the left. All artificial methods of synthesizing organic compounds yield roughly mixtures of D and L molecules. However, living things consist of *either* D or L forms, depending on the species, but never *both*." Even though compounds made in a laboratory from petroleum may be chemically identical atom for atom, they are missing characteristics found in living things.

See also FOOD.

Nuts, Seeds, and Butters

🠅 ORGANICALLY GROWN. Natural food stores and catalogs sell many nuts and seeds in bulk, and nut and seed butters that are organically grown.

🠅 NATURAL. Packaged nuts generally do not contain any additives, but check the label to be sure. Better to buy whole nuts and crack them, or buy nuts and seeds in bulk at the natural food store (they're less expensive there, too). Natural peanut butter and other nut butters are sold through natural food stores and catalogs, but check your supermarket brands as you may find one that is sugar- and additive-free.

See also FOOD.

Oven Cleaners

🠅 PREVENT SPILLS. You'll never have to clean your oven if you cook food in proper-sized containers, or put a cookie sheet on the lower rack to catch spills. If after your preventive measures food does end up at the bottom of the oven, clean it as soon as the oven has cooled to prevent it baking on even more.

🠅 MAKE YOUR OWN. My friend Gina, who is a professional nontoxic housecleaner, uses a mixture of two tablespoons liquid SOAP, two teaspoons borax (in the laundry section of the supermarket), and warm water in a spray bottle. Make sure the salts are completely dissolved to avoid clogging the squirting mechanism. Spray it on, holding the bottle very close to the oven surface so the solution doesn't get into the air (and into your eyes and lungs). Even though these are natural ingredients, this solution is designed to cut heavy-duty oven grime, so wear gloves and glasses or goggles if you have them. Leave the solution on for twenty minutes, then scrub with steel wool and a nonchlorine SCOURING POWDER. Rub baked-on black spots with pumice, available in stick form at hardware stores.

🠅 NATURAL/RENEWABLE, HYBRID-NATURAL INGREDIENTS, NONTOXIC/NONRENEWABLE. I know of no commercial oven cleaners that are better for health or the environment.

🠇 TOXIC/NONRENEWABLE INGREDIENTS. Don't use products that contain lye, ammonia, detergents, or aerosol propellants. Note "DANGER" warning on package label.

See also CLEANING PRODUCTS.

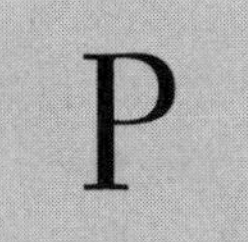

Paint

All paint is made from four categories of ingredients. Solvents, which are usually the largest component found in paint, serve as a carrier to dissolve and disperse the other ingredients. Paints are generally classified according to the type of solvent they contain—oil-based paints contain VOC solvents (40 to 60 percent) and water-based paints use water as the primary solvent (though they still generally contain 5 to 10 percent VOC solvents). The other types of ingredients are resins for adhesion and durability, pigments for color and hiding power, and additives to enhance performance properties.

🠅 MADE FROM NATURAL/RENEWABLE INGREDIENTS. Two German companies sell paints in America made from natural ingredients. Their common philosophy is to use ingredients produced from nature which nature can replenish and will return to nature, and are safe for the user's health. **Livos** paints contain nontoxic petrochemical solvents, and **Auro** paints are completely natural (though they use a renewable balsam terpene solvent that is considered in Europe to be of questionable safety). Despite any small amount of VOCs these paints may emit, these products are much more sustainable than those made from petroleum products.

♠ LOW- OR NO-VOCS. In the Los Angeles basin, famous for its smog, one of the largest sources of VOCs is paints for buildings and other structures. According to the California Air Resources Board, more than half (100 million pounds) of the annual 176 million pounds of VOC emissions generated in California come from just paints and coatings. In the Los Angeles area paints release more VOCs than all of the region's oil refineries and gas stations combined. The EPA has estimated that nationally VOC emissions from architectural paints and coatings exceed 11 billion pounds each year. Southern California has the strictest VOC regulations for paints in the country, requiring emissions of no more than 250 grams per liter.

Oil-based paints contain the highest levels of VOCs. In California you can't even buy them anymore except in very small cans. If oil-based paints are still sold where you live, you can significantly reduce the VOCs you produce by choosing a water-based latex paint, or eliminate them entirely by opting for a paint that emits no VOCs at all.

Both Scientific Certification Systems and Green Seal have standards for paints; however, each has created its standards from a different point of view. SCS measures VOCs that are emitted from the formulated paint, whereas Green Seal addresses the paint ingredients themselves.

SCS has a standard for Volatile Organic Compounds in Architectural Paints and Coatings. Their definition of a VOC is "any organic compound which is emitted into the atmosphere during the use, application, and/or curing/drying of a paint or other architectural coating. The definition includes both VOCs that react with nitrogen oxides in the lower atmosphere to form ozone and VOCs that deplete ozone in the upper atmosphere."

The VOC standard for certification is established as less than one gram per liter of product, a level equaling one tenth of 1 percent—many times lower than the California regulations. If a product meets this standard, it may claim "This product contains no smog-producing ingredients/compounds/chemicals/VOCs, based on a certification standard of one gram per liter." If so qualified, a manufacturer may also claim the product contains no "ozone-depleting ingredients/compounds/chemicals."

Scientific Certification Systems has certified the interior latex paint **Spred 2000,** made by The Glidden Company. When SCS tested Spred 2000 against a certification standard of one gram VOC per liter, no VOCs were detected at that level. The company has set the goal of eliminating petroleum-based solvents from all their decorative architectural paint products by the year 2000. Also certified was their companion brand **Lifemaster 2000** for professional painters.

Green Seal's standards for paint allow a higher emission level of VOCs (fifty to two hundred grams/liter depending on gloss) but specifies a long list of toxic chemical compounds that "shall not be intentionally introduced to the product." Their certification statement would read "This product meets Green Seal environmental standards for reduced toxic chemicals."

A new paint based on English technology is **Ecos Paint,** made from minerals and a vinyl acetate copolymer (derived from nonrenewable petroleum) that is specially treated to eliminate toxicity. The end result is VOC-free and virtually odorless.

🡅 RECYCLED. While most brands of paint you find on the shelf have been formulated from virgin materials, some county household hazardous waste collection programs process and resell their collected paints. Check with your local household hazardous waste program to see if recycled paints are available in your area. Two brands of recycled water-based latex paints are **Cycle II** (which comes in nineteen shades of flat, semigloss, and primer) and **Green Paint** (which even reuses empty paint cans). Though these paints do emit VOCs, they emit no more than regular latex paint, and have the advantage of recycling what would otherwise be household hazardous waste.

Paper

🡅 MAKE YOUR OWN RECYCLED PAPER. Art supply stores carry papermaking kits and instruction books, so you can turn your scrap paper into useful works of art.

🡅 NONTREE FIBERS. More and more we are going to start seeing paper, particularly for personal and business stationery, made from nontree fibers. While you may find handmade papers at a local shop made from almost any kind of cellulosic material, there are three nontree fibers that are being used to produce paper suitable for use as letterhead and for use in laser printers. A soft white paper (and envelopes, too) is made from kenaf, a slender, woody plant that produces up to five times more pulp per acre than trees, with the same strength and performance as wood fiber. A textured wheat-colored paper is being made in India from jute, an ancient vegetable fiber. **Tree-Free Eco Paper** is made from hemp, which produces four times more material for papermaking than trees, combined with cereal straw (a by-product of grain production) to make an off-white vellumlike paper. All these non-tree papers can be ordered from ***Real Goods*** and ***Earth Care.***

◆ RECYCLED PAPER. Some recycled papers you can easily buy include paperboard box packaging (look for boxes that are brown or gray on the inside), bathroom and facial tissues, paper towels and napkins, greeting cards and wrapping paper, office papers such as copy paper and lined tablets, business and personal writing papers, printing papers, and newspapers. I've seen recycled papers in office supply stores and discount office warehouses. Recycled household papers are in every supermarket and natural food store here. Ask your printer to print on recycled paper. If you can't find recycled papers locally, you can order them by mail from ***Atlantic Recycled Paper Company, Brush Dance, Earth Care, Seventh Generation,*** and ***Quill. Lyn-Bar Enterprises*** will do custom printing on 100% post consumer recycled paper with soy-based inks and ship your printing right to your door.

Recycled papers contain both pre-consumer and post-consumer material. Correctly labeled, a product should indicate the percentage of each. If they give the source of the recycled material, even better. An excellent label on **Green Forest Bathroom Tissue** (available in many supermarkets) states "made from 100% recycled paper fibers, including a minimum of 10% post-consumer content. Sources of recycled paper fibers include envelopes, direct mail pieces, printer's overruns, computer printouts, magazines and office paper. Sources of post-consumer paper fibers include similar kinds of paper that have been recovered from consumers' homes and businesses and that have served their intended use. Even the core is made from 100% recycled paper fiber, including a minimum of 10% post-consumer content."

While there are some 100 percent recycled paper products, most are a blend of recycled and virgin pulp to capitalize on the strengths of both. Blends generally contain 40 to 80 percent recycled fibers. Some recycled papers have an off-color, speckled recycled look, but premium grade recycled papers are virtually indistinguishable from comparable grades made from virgin pulp fiber. Recycled papers lie flatter, have strength, shrink and stretch less, and resist moisture and curling, making them in some ways superior to virgin papers for many uses.

In the absence of any other national standard, within the recycled products industry, many have looked to the EPA Guidelines for guidance as to what constitutes products made from certain recycled materials. Many recycled paper labels and swatchbooks state "This paper meets (or exceeds) current EPA Standard 40 CFR Part 250." This doesn't mean there is a regulation required by law, it simply means that the product is eligible for purchase by federal agencies according to the EPA criteria.

The Recycling Advisory Council's recommended content requirements for total recycled fiber and post-consumer recycled fiber, respectively, by product are as follows (these are higher than those recommended in the EPA Guidelines):

PRODUCT	% OF TOTAL RECYCLED CONTENT	% OF POST-CONSUMER CONTENT
Newsprint	40	40
Coated printing and writing paper	40	10
Uncoated printing and writing paper	50	15
Bath, towel, napkin, and wiper tissue	80	70
Facial tissue	60	50
Construction paper	80	65
Paper packaging	40	20
Uncoated paperboard corrugated containers	40	35
Uncoated folding cartons, industrial paperboard, and miscellaneous paperboard products	100	60
Coated folding cartons and miscellaneous coated paperboard products	90	45

Though it is good for the environment to utilize recycled paper in the manufacture of new paper, it is not an environmentally perfect product. Various stages of the recycling process can produce toxic emissions, particularly the deinking and the bleaching processes. Both of these are needed to make pure white paper out of post-consumer paper that has been printed.

The most sustainable recycled paper is unbleached, non-deinked 100 percent recycled paper with a high percentage of post-consumer material. Such a paper, while it already exists, is usually a light gray or tan, with a few "ecology spots." It takes a change in our aesthetic perceptions to accept these. Personally, this is just the kind of paper I use—when I look at it and touch it I am reminded of the environmental good and it becomes beautiful to me. And some recycled papers are beautiful in their own right—their imperfections lend character and interest to the paper.

But this doesn't mean that papers with less recycled content or less post-consumer material are a poor choice. Remember, many pounds of manufacturing waste are created for a pound of consumer product. That manufacturing waste needs to be used, too. And paper fibers do deteriorate after many recyclings, so there will always be a need for virgin stock. Hopefully, in the future, virgin pulp will come from sustainably managed forests instead of clear-cuts.

Both Scientific Certification Systems and Green Seal have standards for recycled printing and writing papers.

Scientific Certification Systems requires that the percentage of recycled post-consumer waste used must be at least 10 percent by weight or greater if required by law in the areas where the product or package is sold. In addition, the total recycled content (pre- and post-consumer material) must fall within a state-of-the-art range (equal to or exceeding 80 percent of the highest percentage of recycled material that can be technically and operationally incorporated into a specific product on a significant scale). The product must also be recyclable, unless it cannot be made into anything else but a nonrecyclable form (such as facial tissues). SCS certified recycled printing and writing papers include **Fox River Circa '83, Circa Select, Confetti, Early American, Fox River Bond,** and **Proterra** papers.

The Green Seal certification has two recycled content standards: (1) a minimum of 50 percent recovered paper (includes pre- and post-consumer material) including at least 10 percent post-consumer material by weight, or (2) a minimum 25 percent post-consumer waste. The recovered material can not be deinked using chlorine or any other EPA-listed toxic solvent, nor can chlorine or any of its derivatives (such as hypochlorite and chlorine dioxide) be used as a bleaching agent in any aspect of the paper production. Levels of lead, cadmium, mercury, or chromium cannot exceed 100 parts per million by weight. Green Seal certified recycled printing and writing papers include **Mohawk Vellum, Mohawk Vellum PC 100, Tomohawk,** and **Irish Linen.**

Virtually every supermarket and natural food store in my area now stocks household papers made from recycled paper—paper napkins, paper towels, toilet paper, and facial tissues. If they're not in your local stores, they will be soon. Some brands have been deinked and are off-white in color; others are not deinked and are decidedly gray.

The Green Seal standards for facial and bathroom tissues require 100 percent recovered paper (including pre- and post-consumer material) with bathroom tissue containing at least 20 percent post-consumer fiber and facial tissue at least 10 percent post-consumer fiber. In addition, if no chlorine is used to bleach or deink the recycled fiber, the manufacturer can also say "No Chlorine Bleach" or "Unbleached" on the label. For companies that do use chlorine for bleaching, Green Seal has limited the levels of chlorine by-products that can flow from the factory. No inks or fragrances are allowed on these products.

The Green Seal proposed certification standard for paper towels and napkins requires 100 percent recovered material including at least 40 percent post-consumer material by weight. Chlorine and other EPA-listed solvents cannot be used, and there can be no dyes, inks, or fragrances (although they may be printed with certain low-toxic inks). In addition, the packages must be sized to promote source reduction.

Scientific Certification Systems certified recycled bath and kitchen paper products include **Aware, Capri, Cascades, C.A.R.E., Envirocare, Enviroquest, Forever Green, Gayety, Gentle Touch, Green Forest, Green Meadow, Nature's Choice, New Day's Choice, Pert, Project Green, Renew, Safe, Today's Choice,** and **Tree-Free** brands. **Aware, C.A.R.E.,** and **Project Green** brands are also certified by Green Seal.

🡅 Reclaimed Cotton. There are also papers made from reclaimed cotton. Cotton produces a paper far superior in quality to wood, and has been used only for fine writing and business correspondence papers. Until recently all paper made by **Crane & Company** was made from postindustrial cotton textile mill and manufacturing cuttings, leftover seed hairs culled from the cotton ball during the ginning of cotton into thread, and cotton linters (the short fuzzy fibers that remain attached to the cotton seed after ginning that are too short to spin), although now they have started to make specialty papers out of post-consumer recycled materials such as old currency. Some **Strathmore** papers are also made of cotton (fine stationery stores carry both brands). As manufacturers have begun to look for new uses for fabric scraps that would have otherwise gone to a landfill, other fabrics are also being made into paper; some of the scrap from manufacturing jeans is being used to make various shades of blue paper and cardboard.

🡇 White, Virgin Paper. Don't use unless you absolutely have to. This is made from wood pulp from clear-cut forests and bleached with chlorine that creates toxic dioxins (which are dumped by manufacturers into local rivers).

Pasta and Sauce

🡅 Organic. Natural food stores and catalogs all have many varieties and brands of pasta and sauces made with organically grown ingredients.

🡅 Natural. Pasta and sauce are two of the few supermarket items that generally do not contain any additives (although some sauces contain sugar).

See also Food.

Pencils

🡅 Refillable. Preserve forests needed for wood pencils by using a refillable pencil that requires only replacement leads.

🡅 Twigs. A novelty item sold in many greenstores and gift shops are pencils made from twigs. These are hand-collected (usually from fallen twigs) and filled with graphite or colored leads.

↑ Unpainted Wood. There are several brands of unpainted pencils. One made by the Blackfoot Indians is sold in many art supply stores, gift shops, and catalogs. **American Naturals** pencils, available almost everywhere office supplies are sold, are "made from 100% sustained-yield cedar. Contains no rain forest wood." They are made from California-grown cedar, which grows interspersed with other species in mixed conifer forests. But their definition of sustained yield is based on massive replanting efforts and is not the same philosophy as has been discussed in this book. Still, it's an improvement over a pencil made from clear-cut rain-forest wood.

↓ Clear-Cut Rain-Forest Wood. Most wooden pencils are made from jelutong, a wood taken from tropical rain forests in Indonesia. Unless the pencil maker labels the type of wood used, it is probably clear-cut jelutong.

Pens and Markers

↑ Fountain Pens. Fountains pens that can be refilled with ink right from the bottle are an elegant and ecological writing implement. Even if you refill with plastic ink cartridges, it's still better than using disposable pens. A variety of fountain pens are sold in fine stationery and art supply stores.

↑ Refillable. Refillable cartridge ballpoint and felt-tip pens are sold in all office supply stores.

↑ Water-Based Inks. A wide variety of disposable pens and markers in many colors and thickness of line come with water-based inks. Some have ball-point tips, others plastic or felt tips. You can tell inks are water-based if they are odorless and flow smoothly. Some markers are labeled "nontoxic."

↓ Toxic Solvent-Based Inks. Don't use pens or markers with solvent-based indelible inks. Though all inks are made from nonrenewable resources, these contain volatile solvents that are harmful to inhale. You can easily tell which markers contain solvents because they give off a strong smell when you open the cap. The ink in solvent-based ballpoint pens has a more subtle odor; you can smell it slightly, but the real clue is that the ink globs around the tip or on the paper.

Pest Controls

All pesticides, whether for home and garden or commercial use, are regulated by the Federal Insecticide, Fungicide, and Rodenticide Act (FIFRA). Originally passed in 1947 to protect farmers from ineffective and dangerous pesticides, amendments in 1972 shifted the emphasis from safeguarding the pesticide user to public health and environmental protection. This act gives federal control to all pesticide applications and regulates both intrastate and interstate marketing and use of pesticides.

FIFRA is intended to insure that the use of a pesticide will not cause "any unreasonable risk to man or the environment, taking into account the economic, social, and environment costs and benefits of the use of any pesticide." Unfortunately, economic and social benefits have taken precedence over environmental and health costs in the past. Since there are safe, workable, and cost-effective alternatives available that make it unnecessary in most cases to use toxic pesticides at all, zero pesticides should be our goal, or at least minimal use of nontoxic pest controls. Recently the EPA took a first step toward encouraging safer pesticides by offering financial incentives and regulatory relief, speeding the time it takes (usually up to seven years) to bring a new pesticide to market.

The act requires registration of all products sold as pesticides, classification of pesticides for general or restricted use based on toxicity, and suspension of those found to cause unreasonable adverse affects on humans or the environment.

Pesticide labels require the toxicity signal words (DANGER: POISON, WARNING, and CAUTION) and precautionary statements, information on proper treatment on poisoning, what types of exposure require medical attention, in what ways the pesticides might be poisonous to humans and domestic animals (though if it says nothing, don't assume it is completely safe), protective clothing or ventilation requirements, and environmental cautions such as if the pesticide will harm birds, fish, or wildlife (often precautions are given such as "Do not use near lakes or streams").

In addition, the common or chemical name of the active ingredients must be listed, with the percentage of active ingredients and inert ingredients. Since October 1988 all EPA List One (toxic) inert ingredients must be labeled with the statement "This product contains the toxic inert ———." Inerts on List One are "of toxicological concern" because they cause cancer, have neurotoxic or other chronic effects, cause adverse reproductive effects, or have a particular toxic effect to the environment. If the name of the inert ingredient does not appear on the label, it might be a "potentially toxic inert with high priority for testing," and inert "of unknown toxicity," or an inert "of minimal concern." While the law does offer some protection, lack of cautionary labeling does not guarantee no toxic inerts are present.

Every pesticide on the market must be registered with the EPA, and this registration number must be on the front panel of the label written as "EPA Registration No. ——." A code for the factory that makes the pesticide (called the "establishment number") must also be on every container. Usually it can be found under the registration number.

Other requirements include the brand name, the common scientific name for the pesticide, the type of formulation (liquid, wettable powders, emulsifiable concentrations, dusts, etc.), child hazard warning ("Keep out of reach of children") on the front label, net contents, directions for use, a reminder that it is a violation of federal law to use a pesticide product in any manner inconsistent with its labeling, and name and address of the manufacturer.

For all the labeling requirements, what pesticide labels do not tell us is what, exactly, the ingredients are. Philip Dickey of the Washington Toxics Coalition offers this sample of a suggested informative label for pesticides. With labels such as these, we know what each product is made from and can find out for ourselves the safety or danger of the ingredients.

INSECTICIDE

Tetramethrin [(1-Cyclohexene-1,2-dicarboximido)methyl 2,2-dimethyl-3-(2-methylpropenyl)cyclopropanecarboxylate] 0.050%

Sumithrin 3-Phenoxybenzyl d-cis and trans 2,2-dimethyl-3-(2-methylpropenyl)cyclopropanecarboxylate 0.096%

Other isomers 0.004%

SOLVENTS

1,1,1-trichloroethane 40.00%

Methylene chloride 20.00%

PROPELLANT

Isobutane 29.85%

This seems simple and straightforward enough. From this label, we can learn that the primary ingredient of this hypothetical pesticide is 1,1,1-trichloroethane, a Class I ozone-depletor, which is probably more harmful to the environment than the active pesticide ingredient.

✦ LIVE AND LET LIVE. What we generally consider pests are living things that are part of the dynamic, balanced ecosystem in which we live. Each species has its own role to play in the web of life, and to be sustainable, we must respect these other species. In nature, species coexist by establishing their territories and honoring those established by others. You can establish your territory with pests by pest-proofing your home.

First, figure out how pests are getting into your home, and do something to keep them out. Fill holes and cracks, put screens on windows, put up chicken-wire barriers or fences.

Then make your home an unpleasant place for pests to be. Take away their food supply by keeping living areas clean—sweep up crumbs, wipe up spills immediately, wash dishes after eating, store food in tightly closed containers, empty garbage and compost scraps frequently. Dry up their water supply—repair leaky faucets, pipes, and clogged drains. And get rid of any clutter they can hide in—clean out closets, the attic, the garage, the basement, and anywhere else unused or infrequently used items pile up.

Before embarking on any action to rid your home of pests, ask yourself if it really is necessary to kill them. A true pest is one that is doing damage or presenting a health threat. If rats are chewing through your electrical wires, that's a very different situation than having a moth resting on your wall. Flies are unsanitary, but a few spiders won't do much harm and actually work beneficially to control other pests.

If you decide you do need to eliminate a pest, try the most natural, least toxic methods of control first. There are a handful of little tips-for-controlling-pests-nontoxically books around, but the one to get is *Common Sense Pest Control*. It's a hefty and expensive book, but an invaluable guide to the most natural means of controlling any household pest, written by experts.

⬆ HOMEMADE. There are many simple things you can do to control pests without buying any special products. You can shoo them away, vacuum them up, put a glass over an insect and slip a card under it and take it outdoors. Herbal repellents work well, too—most pests don't like the smell of bay leaves, cloves, pennyroyal, lavender, or cedar.

⬆ MECHANICAL AND NATURAL. Many nurseries and hardware stores now carry nontoxic, natural, and mechanical alternative pest controls. Flypaper and mousetraps work as well as fly spray and rat poison. "America's Largest Catalog of Poison Free Pest Control Products" is ***Brody Enterprises,*** with all kinds of products for various household pests.

⬇ TOXIC/NONRENEWABLE. Pesticides are among the most toxic products around the home. The immediate health effects from inhaling some common household pesticides during use include nausea, coughing, breathing difficulties, depressions, eye irritation, dizziness, weakness, blurred vision, muscle twitching, and convulsions. Long-term exposure from repeated use can damage the liver, kidneys, and lungs, and can cause paralysis, sterility, suppression of immune function, brain hemorrhages, decreased fertility and sexual function, heart problems, and coma.

The environmental effects of pesticides used on crops have been well publicized, and many of these same dangers apply to household use as well. While it's unlikely that you might kill a songbird while spraying for flies, pesticides are still made from polluting, nonrenewable petrochemicals that produce toxic waste.

See also INSECT REPELLENTS and MOTHBALLS.

Pet Products

Pet products are no different from products for people: They can be produced in a sustainable way that benefits both pet and the earth, or in a way that is harmful to both.

Natural food stores generally carry a full line of natural pet supplies and books on natural pet care. If you have difficulty finding them, you can order by mail from the ***Natural Pet Care Company.***

Pillows

🠅 NATURAL FIBERS. Pillows are available covered with cotton ticking and filled with down or feathers at every department store. Pillows stuffed with cotton, wool, or silk batting can be ordered by mail from ***Chambers*** and ***The Company Store.*** Pillows filled with kapok or buckwheat hulls can be purchased at most futon shops.

🠅 ORGANICALLY GROWN COTTON AND WOOL. Order by mail from ***Jantz Design*** or ***Heart of Vermont*** (these are covered in organic cotton too).

🠅 RECYCLED SYNTHETIC FIBERS. Filled with spun polyester fill made from recycled PET plastic soda bottles, the **Save the Earth Bed Pillow** is certified by Scientific Certification Systems for its recycled plastic fill.

See also TEXTILES.

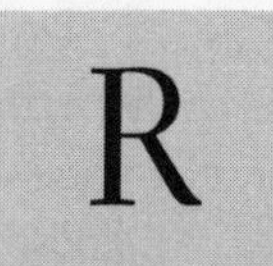

Refrigerators and Freezers

🠅 ENERGY-EFFICIENT. The National Appliance Efficiency Standards specify the maximum electricity consumption of refrigerators according to volume and features. Even so, there is a big difference from model to model, so it pays to shop around. For the most recent evaluation of energy efficient refrigerators and freezers, see the February 1994 issue of *Consumer Reports* magazine. Currently the most efficient models are in the most popular sixteen to twenty cubic foot range; however you might save more energy simply by downsizing. Typically a ten-cubic-foot refrigerator needs only 35 to 60 kWh/month as opposed to a sixteen-cubic-foot model at 100 to 150 kWh/month or a twenty-cubic-foot model at 115 to 180 kWh/month. First pick the right size for you, then look to the EnergyGuide label to find the most efficient model.

Typically, refrigerators last about fifteen years. If yours is getting old, it might be economical both financially and environmentally to replace it. The U.S. Department of Energy has required all refrigerators made after January 1, 1993, to use about 30 percent less energy, and even less in 1998.

Keep in mind that models with the freezer on the top are the most efficient—they use 35 percent less energy than side-by-side models.

🠅 REDUCED CFCs. **Frigidaire** now has a model that advertises "50% Reduced CFCs" (compared to 1987 models) in addition to being 30 percent more energy-efficient than required by federal standards. Look for similar labeling on other brands and watch for new refrigerants to come on the market.

Rug, Carpet, and Upholstery Shampoo

🠅 KEEP YOUR RUGS AND CARPETS CLEAN. Wipe up spills immediately, before they become stains, and vacuum regularly to keep rugs and carpets fresh.

🠅 NATURAL/RENEWABLE OR HYBRID-NATURAL INGREDIENTS. Not yet available to my knowledge, but watch for them.

🠅 NONTOXIC/NONRENEWABLE INGREDIENTS. Not widely available, but a nontoxic carpet shampoo can be ordered by mail from ***AFM Enterprises.***

🠇 TOXIC/NONRENEWABLE INGREDIENTS. Don't use products that contain perchloroethylene, napthalene, ammonia, or detergents. Note warning on package label.

See also CLEANING PRODUCTS.

Rugs and Carpets

🠅 NO CARPETS. The most sustainable choice is to not have carpets at all. More durable are tile, solid hardwood, brick, marble or other stone tiles, or natural linoleum.

🠅 NATURAL FIBERS. Look for area rugs made of cotton, cotton-wool blend, or unmothproofed wool without jute or latex backing in your local import and department stores (sometimes they are in the bath department, but they can be used in other rooms as well). Natural fiber wall-to-wall carpeting is rare in stores, but can be ordered from ***Hendricksen's Natürlich,*** a family-owned company that specializes in natural floor coverings of all kinds.

🡅 RECYCLED SYNTHETIC FIBERS. Ask your local carpet dealer for **Image Carpets,** made from 100 percent recycled PET bottles.

🡅 VAPOR BARRIER. If you have a synthetic carpet and you suspect there is a problem with chemical offgassing (see below), you can apply a nontoxic vapor barrier sealant called **AFM Carpet Guard.** It's completely odorless, but effectively blocks carpet odors.

🡇 SYNTHETIC FIBERS. Rugs and carpets made from synthetic fibers are made from nonrenewable resources and will not biodegrade.

In addition, they can contain up to 120 toxic and carcinogenic chemicals (including installation materials). Symptoms reported to the Consumer Product Safety Commission (CPSC) as being from exposure to synthetic carpet include such diverse problems as burning eyes, memory problems, chills and fever, sore throats, joint pain, chest tightness, cough, numbness, nausea, dizziness, light-headedness, nervousness, depression, and difficulty concentrating. In 1991 the attorneys general of twenty-six states petitioned the CPSC to require health warnings on new carpet and installation materials, but the CPSC refused, saying it was "premature."

The EPA considers synthetic carpet to be a major contributor of VOCs to indoor air pollution and has safety information available for consumers.

Scouring Powders

🡅 BAKING SODA. Pour some in a waterproof container (such as a metal grated cheese shaker) and keep it by the sink. Works great.

🡅 NATURAL/RENEWABLE OR HYBRID-NATURAL INGREDIENTS. The most widely available are the **Bon Ami** products. **Bon Ami Cleaning Powder** and **Bon Ami Cleaning Cake** (available in many hardware stores, greenstores, and green catalogs) are made from nothing more than soap and ground feldspar. **Bon Ami Polishing Cleanser** (available in every supermarket) contains added nontoxic detergents and nonchlorine bleach.

🡇 TOXIC/NONRENEWABLE INGREDIENTS. Don't use products that contain chlorine and detergents.

See also CLEANING PRODUCTS.

Shampoo and Conditioner

🡅 ORGANICALLY GROWN INGREDIENTS. A few brands of shampoo and conditioner at your natural food store contain organically grown or biodynamically grown herbs, including **Weleda Shampoos** (you can also order from ***Meadowbrook Herb Garden***).

🡅 NATURAL/RENEWABLE AND HYBRID-NATURAL INGREDIENTS. Most shampoos and conditioners you'll find at the natural food store fall into this category.

See also BEAUTY AND HYGIENE.

Shaving Products

🡅 REFILLABLE RAZOR. Instead of using disposable razors or an electric razor (though it has been argued to me that an electric razor could be sustainable if it was powered by the sun), use a refillable razor.

🡅 ORGANICALLY GROWN INGREDIENTS. A couple of shaving products at your natural food store will have some organically grown or biodynamically grown herbs. **Speick Shaving Products** contain biodynamically grown herbal extracts (order from ***Meadowbrook Herb Garden***).

🡅 NATURAL/RENEWABLE AND HYBRID-NATURAL INGREDIENTS. Your natural food store should have a good selection of these.

See also BEAUTY AND HYGIENE.

Shoe Polish

🡅 WEAR SUEDE OR FABRIC SHOES. These do not require polishing.

🡅 NATURAL/RENEWABLE OR HYBRID-NATURAL INGREDIENTS. **Livos** shoe polish is available in some greenstores and green catalogs, or by mail directly from the importer ***Natural Choice***.

↑ NONTOXIC/NONRENEWABLE INGREDIENTS. A shoe polish free from petroleum distillates is made by **AFM.**

↓ TOXIC/NONRENEWABLE INGREDIENTS. Don't use products that contain aerosol propellants or toxic solvents such as methylene chloride, nitrobenzene, perchloroethylene, trichloroethane, trichloroethylene, and xylene.

See also CLEANING PRODUCTS.

Shoes and Slippers

↑ NATURAL FIBERS. A couple of classic shoe styles are made from cotton. Authentic espadrilles are made from cotton and twine; tennis shoes at least have cotton uppers; and cotton maryjanes are imported from China.

↑ LEATHER. All-leather shoes are available at both ends of the price spectrum—either they are very expensive fashion shoes or they are very inexpensive imports from Third World countries sold at import stores.

↑ RECYCLED MATERIALS. The most environmentally responsible shoes I know of are **Eco Sneaks** and **Envirolite** shoes for men and women (sold in many greenstores and environmental catalogs). They are made from a myriad of recycled materials (the complete list is given in Chapter 6).

↓ SYNTHETIC MATERIALS. Shoes made from synthetic materials use nonrenewable resources, are not biodegradable, and can cause foot rashes because synthetic materials won't let the shoe breathe.

See also TEXTILES.

Shower Curtains

↑ GLASS ENCLOSURE. Eliminate the need for a shower curtain by installing glass doors.

↑ COTTON. I'm starting to see cotton shower curtains being sold all over—even in discount bedding stores. It is generally recommended that you use a plastic liner with the fancy ones; however, you can use a plain cotton duck shower curtain by itself. If you can't find one locally, order from ***Clothcrafters*** or ***Vermont Country Store.***

🡇 SYNTHETIC MATERIALS. Don't use. Shower curtains made from vinyl plastic or nylon use nonrenewable resources and are not biodegradable,

Shower Heads

🡅 LOW-FLOW. A low-flow shower head can reduce the amount of water used from five to seven gallons per minute (gpm) to about three gpm (40 to 60 percent reduction)—and still provide a satisfying shower. Green Seal has a certification standard of 2.5 gpm. If you take a ten-minute shower, that's a savings of about thirty gallons each shower, or 10,950 gallons per year. Unless you know you have a low-flow shower head, you probably have one that uses a lot of water, especially if you have an older model.

There are two types of low-flow shower heads. Aerated shower heads are the most popular. They reduce the amount of water in the flow, but maintain pressure by mixing air in with the water. Just like a regular shower head, they produce a steady, even spray. Many also have on-off levers that make it easy to restrict the flow while you're soaping up, then allow the water to come through full force when you want to rinse off. The nonaerated variety simply reduces the flow. It gives a good, forceful spray, but the flow pulses like a massage shower head.

Silver Polish

🡅 MAKE YOUR OWN. The best way I know of to clean silver is to magnetize the tarnish away. The basic ingredients needed are aluminum (in the form of either a pot, pan, or aluminum foil) and some kind of salt (table salt, rock salt, sea salt, and baking soda all work fine). In salty water the aluminum will act as a magnet and attract the tarnish away from the silver. After submerging the pieces of silver for a few minutes in water containing both the aluminum and the salt, you can literally wipe them dry and the tarnish will be gone (badly tarnished silver may need to go through the process several times).

🡅 NATURAL/RENEWABLE, HYBRID-NATURAL, NONTOXIC/NONRENEWABLE INGREDIENTS. I know of no commercial silver polishes that are better for health or the environment.

🡇 TOXIC/NONRENEWABLE INGREDIENTS. Don't use products that contain ammonia or petroleum distillates. Note "DANGER" warning on package label.

See also CLEANING PRODUCTS.

Skin Care

▲ ORGANICALLY GROWN INGREDIENTS. There are a handful of very pure skin-care lines that use some organically grown or biodynamically grown ingredients.

Alexandra Avery products (at your natural food store and by mail) contain biodynamically grown herbs and organically grown oils. All the ingredients are whole and natural—no derivatives are used and all scents, colors, and preservatives come from the ingredients themselves.

Dry Creek Herb Farm mail orders simple skin-care products made with organically grown and wildcrafted herbs.

Dr. Hauschka Skin-Care System is sold in natural food stores (or order by mail from ***Meadowbrook Herb Garden***) and has been on the market for thirty-five years. All plants used are biodynamically grown, and because of special processing they do not need preservatives. All ingredients are active; no extraneous colors or fragrances are added.

A few other natural food store brands also contain some organic herbs along with natural ingredients—read the labels.

▲ NATURAL/RENEWABLE AND HYBRID-NATURAL INGREDIENTS. You'll find many skin-care products at your natural food store.

▼ HARMFUL INGREDIENTS. Avoid products that contain artificial colors, artificial fragrances, mineral oil, preservatives, and other ingredients made from nonrenewable petro-chemicals.

See also BEAUTY AND HYGIENE.

Skin Lotions, Creams, and Moisturizers

▲ ORGANICALLY GROWN INGREDIENTS. A few natural food store brands—including ***Alexandra Avery*** and **Weleda** (order from ***Meadowbrook Herb Garden***)—contain organically grown or biodynamically grown herbs.

▲ NATURAL/RENEWABLE AND HYBRID-NATURAL INGREDIENTS. Your natural food store will have a shelf full of natural lotions, creams, and moisturizers.

See also BEAUTY AND HYGIENE.

Snack Foods

▲ DON'T SNACK ON JUNK FOODS. I know this is an unrealistic suggestion—we're all going to have an occasional chocolate chip cookie or potato chip—but I had to put it in. The healthiest way to eat is to eat good, nutritious food at meal and snack times and not munch idly through the day on empty calories. Unless you're munching on organically grown vegetables and fruits, you're probably eating highly processed, packaged foods that are unnecessary luxuries above and beyond the food you need for sustenance. Take a look at your snacking habits and see if there are some snacks you (and your waistline) could just as well do without. If you eat nourishing food at mealtimes, you shouldn't want to snack.

◆ MAKE YOUR OWN. If you're going to eat snack foods, fix them yourself instead of buying an instant snack in a package. Pop popcorn, bake cookies, crisp tortillas in the oven. They'll taste much better than the packaged foods.

◆ ORGANICALLY GROWN INGREDIENTS. If you're going to buy packaged snack foods, at least choose those made at least in part from organically grown ingredients. There are plenty at your natural food store to choose from.

◆ NATURAL. All snack foods sold in natural food stores are additive-free. Many sold in supermarkets are as well, but check the labels carefully.

See also FOOD.

Soap

Soap is made from animal or vegetable fat and an alkali such as sodium hydroxide or ashes. It has been used for centuries and is absolutely safe and biodegradable under the proper conditions.

Soap is used for cleaning and for personal hygiene. A single soap could fulfill both functions, or you can choose two soaps with different characteristics for each purpose.

As a chemical-engineering textbook from the sixties states, "There is absolutely no reason why old-fashioned soap cannot be used for most household and commercial cleaning." A WATER SOFTENER will improve the performance of soap in hard water, eliminate soap scum, and allow you to use less soap to do the same cleaning job.

◆ MAKE YOUR OWN. You could certainly make your own soap once a year—it's easy enough and a nice weekend activity. Your local library should have instructions.

⬆ ORGANICALLY GROWN NATURAL/RENEWABLE INGREDIENTS. Some personal care soaps at your natural food store will contain organically grown or biodynamically grown herbs.

⬆ NATURAL/RENEWABLE OR HYBRID-NATURAL INGREDIENTS. Your natural food store has a full range of soaps made from plant and animal fats. One brand of liquid soap that has been sold in every natural food store for years is **Dr. Bronner's Pure Castile Soap,** which can be used for cleaning houses and people. It comes in natural fragrances like peppermint, almond, and lavender. ***Cal Ben Soap Company*** has an almond-scented bar soap and a full line of soap-based personal care and cleaning products.

See also BEAUTY AND HYGIENE and CLEANING PRODUCTS.

Sponges

⬆ BIODEGRADABLE NATURAL SEA SPONGES. Sold in most natural food stores and bath shops, these large, soft corals are usually sustainably harvested from the ocean.

⬆ BIODEGRADABLE NATURAL CELLULOSE SPONGES. Every hardware and paint store sells beige-colored sponges in a variety of sizes made from processed natural cellulose.

⬆ BIODEGRADABLE NATURAL LOOFAHS. Most natural food stores and bath shops sell loofahs—long, stiff sponges made from gourds. They are naturally a light golden beige in color; avoid those that are dyed bright colors. To make little scouring pads, cut loofahs into one-inch slices.

⬇ PLASTIC SPONGES. Don't use—these are made from nonrenewable ingredients and are not biodegradable.

Sun Protection

⬆ PROTECTIVE CLOTHING. The most prudent and resource-efficient way to shield your skin from the sun's harmful rays is to wear protective clothing such as windbreakers, shirts, scarves, gloves, and wide-brimmed hats. ***SelfCare Catalog*** has some special sun protection clothing and accessories that would be worth looking into if you're out in the sun a lot.

♠ ORGANICALLY GROWN INGREDIENTS. There is one sunscreen product—**Alexandra Avery Sun Oil**—that is made with oil from organically grown plants and biodynamically grown herbs. A few other natural food store brands contain organically grown or biodynamically grown herbs—check the labels.

♠ NATURAL/RENEWABLE AND HYBRID-NATURAL INGREDIENTS. Your natural food store carries a number of natural sun lotions.

See also BEAUTY AND HYGIENE.

Sweeteners

♠ HONEY. Nature's sweetener, it can be eaten without processing, direct from the beehive. Honey is always pure, because bees exposed to pesticides usually don't make it back to the hive. Legally, for a bottle to be labeled honey, it must contain "the nectar and floral exudations of plants gathered and stored in the comb of honeybees." Honey is always honey, regardless of how it is labeled. U.S. Grade A or Fancy refers to the level of filtration and does not give any indication of quality or processing. There are many varieties and flavors of honey, sold in every kind of food store. Choose a brand made close to home, as you will then be eating honey made from the nectar of your local flora.

♠ MAPLE SYRUP AND SUGAR. One technological step beyond honey, maple syrup is simply boiled down maple sap. Though our bodies respond to its sucrose in the same way as white sugar, maple syrup and sugar is closer to its natural state than any other sucrose sweetener. Maple trees are grown without fertilizers or pesticide sprays, though some American producers inject the trees with formaldehyde pellets to increase the flow of sap (if they do, it's not on the label). Most brands in natural food stores are formaldehyde-free, but this might not be so for maple syrups purchased elsewhere. Check with the producer to verify their sugaring practices.

♠ ORGANICALLY GROWN BARLEY MALT SYRUP AND RICE SYRUP. Barley malt is made by soaking and sprouting barley. It is then combined with more barley to make barley malt syrup or with rice to make rice syrup, then cooked until the starch is converted to sugar. The mash is then strained and cooked down to syrup. These syrups have a gentle character, less sweet and flavorful than honey. Sold in natural food stores and catalogs.

♠ ORGANICALLY GROWN SUGAR. If you want the sweetness and properties of regular white sugar, try **Sucanat Organic Sugar.** This evaporated sugar cane juice is sold in natural food stores and catalogs.

🡅 NATURAL DATE SUGAR. This sweetener is made by dehydrating dates to 2 percent moisture and grinding them very finely into granules. Can be substituted for brown sugar, cup for cup. Sold in natural food stores and catalogs.

🡅 NATURAL FRUIT SYRUP. A thick syrup made from a blend of fruits that are concentrated by removing water, natural fruit acids, and strong flavors. It can be used like honey. Sold in natural food stores.

🡅 NATURAL FIG SYRUP. Made from figs and water, this works very well as a substitute for molasses, cup for cup. Sold in natural food stores.

🡇 FRUCTOSE. Often used as a substitute for sucrose in natural products, fructose is generally manufactured from corn syrup, not fruit. Highly refined and often contains more than half sucrose.

🡇 SUGAR. Sugar has been linked to many health problems, including tooth decay, diabetes, hypoglycemia, coronary disease, obesity, ulcers, high blood pressure, osteoporosis, and nutritional deficiencies. It has many names—Barbados molasses and sugar, blackstrap molasses, brown sugar, cane sugar and syrup, corn syrup, dextrose, invert sugar and syrup, muscavado sugar, raw sugar, ribbon cane syrup, sorghum syrup, turbinado sugar—and comes from many sources, but they all are basically sucrose. Many pesticides are used to grow sugars, and they are highly refined.

🡇 ARTIFICIAL SWEETENERS. Saccharin is still used in some products (such as dehydrated iced tea mix), even though a label must appear warning that saccharin "has been determined to cause cancer in laboratory animals."

The ubiquitous artificial sweetener now is aspartame (sold as NutraSweet and Equal). It's a so-called "natural" sweetener made of phenylalanine and aspartic acid, containing "nothing artificial." In your body these naturally occurring substances break down into the same amino acids found in any protein food. But nature has designed these amino acids to be ingested in balanced amounts along with the many other vitamins, minerals, and enzymes, not to be taken in a concentrated form in excessive amounts. A little aspartame won't hurt, but with it now being used in every food product from breakfast cereals to vitamin pills, it's easy to ingest too much. A variety of symptoms have been associated with overconsumption of aspartame. Scientific tests performed on humans to establish the safety of aspartame prior to FDA approval resulted in depression, menstrual irregularities, headaches, constipation, and tiredness.

See also FOOD.

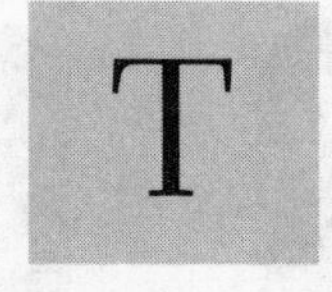

Table Links

Table Linens

♦ NATURAL FIBERS. Choose cotton or linen tablecloths, place mats, and napkins. Import stores often carry colorful, inexpensive styles; department stores or linen shops have fine, often imported table linens. Also check antique shops for heirloom-quality tablecloths and napkins.

See also TEXTILES.

Tea

♦ ORGANICALLY GROWN. There are many varieties of organically grown herbal and black teas sold in natural food stores, both boxed and in bulk. You can order biodynamically grown teas from ***Meadowbrook Herb Garden.*** Loose tea is generally fresher and has a better flavor than tea in bags because it is in its more natural leaf or flower form, rather than being ground into tiny particles. Besides, it saves resources to buy loose tea; it's easy to brew it using metal tea balls, bamboo strainers, or reusable cotton tea bags.

♦ NATURAL. Many natural herbal and black teas can be purchased in bulk at natural food stores and coffee specialty shops.

See also FOOD.

Textiles

The Textile Fiber Products Identification Act, passed in 1960, requires that each textile product be labeled with the generic names of the fibers from which it is made. The generic names are established by the Federal Trade Commission and include twenty-one man-made fiber groups, plus the natural fibers. This law applies to all yarns, fabrics, household textile articles, and wearing apparel, so it's easy to identify the fiber content of textile products we use every day. Imported goods must also adhere to this law, and the label must also reveal the country of origin. This labeling appears along with the name of the manufacturer, the size of the garment, and cleaning instructions, and is usually attached to the seam of a garment near the waist or hemline.

Label information applies only to the fibers used in the body of the garment or item. Sometimes labels will read "100 percent natural fiber, exclusive of dec-

oration," without revealing the fabric of the decoration. Sweaters labeled "100 percent cotton" often have nylon threads running through the bottom edge and sleeve cuffs to help retain their shape. Cotton chamois shirts sometimes have nylon interfacings behind the buttons. Polyester thread may be used in natural fiber garments, as well as synthetic zippers, elastic, trims, linings, and interfacings, and plastic buttons and hooks.

Exempt from this regulation are upholstery stuffing; outer coverings of furniture, mattresses, and box springs; linings, interlinings, stiffenings, or paddings incorporated for structural purposes and not for warmth; sewing and handicraft threads; and bandages and surgical dressings.

In addition to government regulations, independent organizations sell logos to manufacturers, giving information about the fiber content or performance of the textile products that bear them. The use of these logos is regulated, and products are tested to make sure they live up to their claims (see below).

🡅 NATURAL FIBERS. All the natural fibers are renewable resources, and have the potential to be sustainable. The fiber itself is fine—what needs to be improved are the growing and processing practices, which are highly chemical-intensive and polluting. Still, natural fibers are renewable and biodegradable; synthetic fibers aren't.

Cotton comes from the fibers that develop around the seedpod of the cotton plant. Sanforized cotton has been precompressed by a mechanical process to the size to which it would shrink after washing. Mercerized cotton has been immersed under tension in a strong solution of lye to improve the strength, absorbency, and appearance of the fabric and make it colorfast. Cotton is also often combined with other natural fibers: cotton-silk, linen-cotton, and wool-cotton (commonly known as Viyella).

The Seal of Cotton is a trademark of Cotton Incorporated, and can be used only with its permission. Created in 1973 this brown-and-white logo indicates that the item is made from 100 percent domestic cotton. It is used on fabrics, garments, bed linens, and towels. Be careful to not confuse it with the *Natural Blend* seal. In use since 1974, and also a trademark of Cotton Incorporated, it can be used on fabrics containing up to 40 percent synthetic fiber.

Linen fibers come from the inner bark of the flax plant and were possibly the first fibers used by man. Linen is often used in its natural beige shade, but is also bleached or dyed. Linen is sometime combined with cotton or silk; in colonial times it was generally mixed with wool to make linsey-woolsey.

Ramie is a stingless nettle indigenous to mainland China. Most ramie is grown in Asia, and is sold in America primarily blended with cotton in sweaters.

Rayon is a quasi-natural fiber. Although rayon is a man-made fiber, it is composed of cellulose, a substance found in all plants, instead of pure petrochemicals. Cellulose used in making rayon is taken from cotton linters, old cotton rags, paper, and wood pulp. The cellulose is broken down with petrochemicals and then reformed into threads resembling cotton or silk; however, the process is very toxic. A similar fabric is the new Tencel, which is also made from cellulose using a more environmentally benign process.

Silk is a protein fiber taken from the cocoon of the silkworm caterpillar. Each cocoon is spun from one continuous silk filament extruded from the caterpillar's body.

Wool is a generic term that refers to protein fibers spun from the fleece of more than two hundred different breeds of sheep, and from the hair of angora rabbits, cashmere goats, camels, alpacas, llamas, and wild vicuñas.

The Woolmark is a registered certification owned by the Wool Bureau, Inc., a division of the International Wool Secretariat. Labels are purchased by manufacturers for apparel, knitwear, floor coverings, upholstery materials, blankets, and bedspreads. Since 1964 more than fourteen thousand companies in more than fifty countries have been licensed to use it. Items bearing this logo must be of pure wool and be labeled "100 percent pure wool," "100 percent virgin wool," "all pure wool," or "all virgin wool." Do not confuse it with the Woolblend mark, which was introduced in 1972 by the Wool Bureau, Inc. Products displaying this logo can contain up to 40 percent other fibers (all nonwool components by their generic names).

Leather is generally the skins of animals that have been slaughtered for meat.

Down and feathers are taken from ducks and geese.

Kapok is a fiber taken from the seedpod of the tropical kapok, or silk-cotton tree. It is used as a natural stuffing material.

🡅 ORGANICALLY GROWN COTTON. I first became aware of the organic claim on cotton ten years ago, when only a few acres were being grown as an experiment in Texas. There were no certifications then, and, as I have since found out from the Texas Department of Agriculture, neither was this cotton organically grown. No defoliants were used on this particular cotton, but it was grown with pesticides and artificial fertilizers.

Because of the increasing numbers of organic cotton claims, the Texas Department of Agriculture began to certify organically grown cotton in 1991 and now has a full-scale program. They certify all phases of production from the

The sewn-in Woolmark label is your assurance of quality-tested fabrics made of the world's best...Pure Wool.

The sewn-in Woolblend Mark label is your assurance of quality-tested fabrics made predominantly of wool.

growing through the ginning process. Textiles made from organically grown cotton often have improvements in other aspects of their production as well, such as the use of less toxic dyes, and the elimination of finishes.

The most innovative organic cotton by far is **FoxFibre,** developed by Sally Fox. In 1982, while trying to breed insect resistance into cotton plants, she noticed that occasionally a cotton plant produced green or brown cotton, just as occasionally a flock of white sheep has a few black lambs. So far she has commercialized green and brown cottons, and says that light pinks and blues will probably be possible after years more breeding work. The colors deepen with age rather than fading, as dyed fabrics do. While Sally's breeding nursery is certified organic, unfortunately, not all the colored cotton grown on contract is grown using organic methods; pesticides are used, but no defoliants. If the label says **FoxFibre Colorganics,** the fiber has been organically grown.

🠅 "GREEN COTTON." Though not organically grown, "green cotton" (as it is often called) has no dyes, bleaches, or formaldehyde finishes. There has been a rumor going around that bleaching cotton creates toxic dioxins in a similar way to bleaching paper; however, I have been unable to confirm this. I doubt that this is the case, as the dioxins in paper are created by the interaction of chlorine with the lignin in wood, an organic substance that is not present in cotton or other fibers. I have also heard that many bleached cottons are bleached with hydrogen peroxide instead of chlorine, so there is no formation of dioxin. Regardless of the dioxin question, chlorine is a toxic, ozone-depleting substance which must be manufactured and disposed of, so there is some environmental benefit to not using it. These unbleached cottons can be trusted to be formaldehyde-free, as these fabrics are diverted from the manufacturing process before dyes or finishes are applied.

A good source for green cotton items is the ***Seventh Generation*** catalog.

They have a fine selection of clothing for the whole family, bedsheets, comforters, blankets, kitchen and table linens, towels, diaper-service quality diapers, baby receiving blankets, and a "string bag" baby carrier/sling.

🡅 Recycled Natural Fibers. These are made from pre- and post-consumer wastes, which are collected, unraveled, then respun into yarn for weaving or knitting.

According to the Council for Textile Recycling, nationwide more than 4 million tons of post-consumer textiles enter the waste stream every year. Most go to landfills, but around a million tons are collected by charity groups. About half the textile products collected are sold as secondhand items, and the rest goes eventually to rag graders. There they are sorted for different markets and either sent to other countries for sale as used clothing, or chopped up and used to make items such as blankets.

In addition to our post-consumer fabric waste, 25,000 tons of new textile fiber is disposed of by North American spinning mills, weavers, and fabric manufacturers each year. **Eco Fibre Canada** has come up with a patented (pending) process to make new cotton yarn from gin waste, commercial fabric trimmings, mills ends, and used clothing. So far the company has several recycled cotton fabrics that differ in size of thread, fiber content, percentage recycled fiber, waste source, and other specifications. Eventually such recycled cotton fabric may be sold in every store.

Recycled or respun cloth is not yet common, but is starting to be seen in a few products.

🡅 Recycled Synthetic Fibers. Some polyester fabrics and stuffing materials are now being made from recycled PET plastic soda bottles.

🡇 Fabric Finishes. Because cotton wrinkles easily, many cotton fabrics are treated with a formaldehyde-based resin for easy care. Labels on fabrics so treated usually say "permanent pressed," "no-iron," or "crease resistant" because it is a selling point to the consumer, not because it is required by law. These finishes combine formaldehyde resin directly with the fiber, making the formaldehyde irremovable. Symptoms from inhalation of vapors include coughing, swelling and irritation of the throat, watery eyes, respiratory problems, headaches, rashes, tiredness, excessive thirst, nausea, disorientation, asthma attacks, and insomnia.

🡇 Dyes. Dyes can also be a problem. Usually dyes are colorfast, but some dyes do bleed: we've all ended up with colored armpits or with some white clothes that turned pink because we accidentally washed them with red clothes. If the dye comes out in water, it can be absorbed by the skin. Virtually all commercial dyes are made from nonrenewable, polluting petrochemicals, and many use heavy metals which are dumped as toxic waste.

🡇 NONRENEWABLE SYNTHETIC FIBERS. Don't use synthetic fibers such as polyester, acrylic, and nylon. There is little scientific evidence to prove that these fibers are harmful to health, but environmentally these fibers do more harm than good—they are made of nonrenewable crude oil, they produce toxic waste in their manufacture, they don't biodegrade, and they can't be recycled.

See also BATH LINENS, BED LINENS, BEDS, BLANKETS AND AFGHANS, CLOTHING, COMFORTERS AND QUILTS, FABRIC, KITCHEN LINENS, PILLOWS, SHOES AND SLIPPERS, SHOWER CURTAINS, and TABLE LINENS.

Toilets

🡅 COMPOSTING. The ultimate water-saving toilet is a composting toilet. They require little or no water and leave you with rich compost for your garden (there is even an earthworm-composting toilet that works by vacuum flush and eliminates potential odor and insect problems). Rocky Mountain Institute keeps an up-to-date listing of what's available in composting toilets. Check local building codes first, though, as composting toilets are illegal in some areas.

🡅 LOW-FLOW AND ULTRA-LOW-FLOW. A typical toilet will use five to eight gallons per flush (gpf), whereas ultra-low-flush toilets use only about 1.5 gpf (a 70 to 80 percent savings). Ultra-low-flush toilets work as well or even better than low-flush models (which use 1.6 to 3.5 gpf), and can cost as little as one to two hundred dollars. Green Seal has a standard for certification of a 1.6 gallon maximum flush volume.

In the past low-flow flush toilets have been notorious for not flushing, but recently, because of the great interest in water saving, technology has improved. Some newer models have a turboflush mechanism that uses the pressure from the water pipe instead of gravity, so when you flush, the water moves with a big whooooosh, and everything moves out of the toilet bowl while using only 1.5 gallons of water. All the leading toilet manufacturers have this technology, and low-flush toilets are now sold in most hardware and home building stores, so there's no reason to continue to waste water with an old toilet.

🡅 RESTRICTORS. If you can't or don't want to buy a new toilet, you can reduce the amount of water used by purchasing toilet dams or other gadgets designed to restrict the amount of water in the tank. There is even a basin that sits on top of the toilet that reroutes clean water to the basin when you flush so you can wash your hands, then filters the wash water into the tank and bowl where it is used for the next flush. Whatever you choose, check for the actual gallons of water saved, and don't just rely on the claim that the product "saves water."

You can order water-saving toilets and devices by mail from ***Real Goods.***

Toothpaste and Toothbrushes

⬆ MAKE YOUR OWN. Use plain baking soda or mix baking soda with a few drops of peppermint, anise, or cinnamon essential oil. If allergic to these oils, do not use.

⬆ NATURAL/RENEWABLE AND HYBRID-NATURAL INGREDIENTS. There is no shortage of natural toothpastes at your natural food store. **Tom's of Maine Toothpaste** can also be found in many drugstores and supermarkets.

⬆ WOODEN TOOTHBRUSH WITH NATURAL BRISTLES. Renewable and biodegradable. If your natural food store doesn't carry the **Brooks Pearwood Toothbrush,** order by mail from ***Traditional Products Company*** or ***Heart of Vermont.***

⬆ PLASTIC TOOTHBRUSHES WITH REPLACEABLE HEADS. Even though they are made from nonrenewable, nonbiodegradable plastic, these toothbrushes reduce some plastic waste. Because the bristles wear out faster than the handle, these toothbrushes come with several replaceable heads. Sold in many catalogs and natural food stores.

⬇ PLASTIC TOOTHBRUSH. Made from nonrenewable resources and will not biodegrade.

See also BEAUTY AND HYGIENE.

Trash Bags

⬆ RECYCLED PAPER. These are used primarily for biodegradable garden wastes. Look for them in stores with garden centers.

⬆ RECYCLED PLASTIC. Scientific Certification Systems has certified **Best Buy, Full Circle,** and **Renew** brand trash bags for recycled plastic content. They are available at most supermarkets.

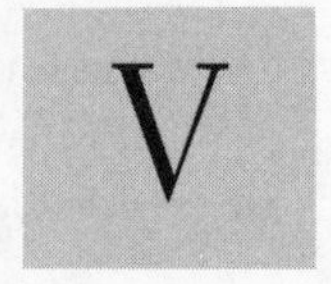

Vinegar

🠉 ORGANICALLY GROWN INGREDIENTS. Your natural food store should have vinegars made from organically grown ingredients. Organic apple cider vinegar can be ordered from ***Walnut Acres;*** organic brown rice vinegar from ***Gold Mine Natural Food Company.***

🠉 NATURAL. All plain vinegars—apple cider, white, wine, balsamic, rice—are additive-free, though some of the flavored specialty vinegars may contain additives. Read labels carefully.

See also FOOD.

Water for Drinking

🠉 PURE SPRING WATER FROM NATURE. Though this is very rare nowadays, I have to mention it because it is the most sustainable option, and available in some places. I had the immense pleasure once of drinking uncontaminated water directly from a spring fed by snowmelt in the town of Mt. Shasta, California. The spring is at the edge of the town and people fill their own bottles to use this water for drinking. Make sure that any spring water you drink is free from contaminants.

🠉 BOTTLED WATER IN RECYCLABLE GLASS BOTTLES. There are many good bottled waters on the market. These are closer to their natural state, even though processed during bottling, than water that has been contaminated, then purified. Choose a brand that is bottled close to where you live, rather than one that is imported.

Bottled water is defined by the Food and Drug Administration as simply "water that is sealed in bottles or other containers and intended for human consumption." Many people believe that bottled waters are of higher quality than tap water, although legally this need not be true. Legally accepted sources for bottled water are wells, springs, and public water right from the tap. No requirements specify that the source of the water or any treatment it has undergone be listed on the label, but if any information is given, it must be truthful. So look for labels that say "spring water" or "natural spring water" (not "spring-fresh," "spring-like," or "spring-pure") and avoid those that say "drinking water," which might be just from the tap, or run through some kind of purification device.

The drawbacks to bottled water are the energy and pollution created by the bottling plant and transportation, and the bottles. In addition high levels of pollutants are allowed (though not necessarily present) in bottled waters because of an industry-wide assumption that bottled waters are consumed as a recreational beverage—a health-conscious alternative to an alcoholic beverage or soda pop—and not as the primary source of drinking water. If you do choose bottled water as your primary source of drinking water, find out what's in your favorite brand. Most companies have a water-analysis report available for inspection that they should send on request.

🡅 USE WATER PURIFICATION DEVICES. The safest bet is to purify your own water at home. Even though the water is far from its original source, you have greatest control over how pollution-free it is.

Water Heaters

🡅 SOLAR. These have been around for a long time—at the turn of the century they were common equipment in homes in the sunnier parts of this country. Technology has, of course, improved since then, but the basic idea is still the same. Solar water heaters are also available from the catalogs listed under ENERGY, RENEWABLE.

🡅 TANKLESS. You can reduce your energy consumption for water heating 50 percent (or more if you have an electric water heater) by replacing your tank-style water heater with a "tankless" or "instant" water heater. These are about one fourth the size of conventional systems and heat only the water that passes through them when you turn on the hot water tap. As an added bonus, you never run out of hot water because there is no holding tank (though they do have a limited rate at which they heat water). They come in different sizes, from whole-house to a tiny one that fits perfectly under your sink for an instant cup of hot tea. The most efficient brands and models are listed in *Consumer Guide to Home Energy Savings*. If you can't find instant water heaters at your local home improvement center, you can order them by mail from ***Real Goods.***

🡅 ENERGY-EFFICIENT. A typical water heater has a large storage tank filled with water that must be reheated again and again throughout the day and night to maintain it's temperature, whether you use it or not. Not the most efficient use of energy, but it's our current standard method for heating water.

The energy efficiency of a storage water heater is indicated by its energy factor (EF), an overall efficiency rating based on the use of sixty-four gallons of hot water per day. The National Appliance Efficiency Standards require these minimum EFs:

TANK SIZE	GAS	OIL	ELECTRIC
30 gallons	0.56	0.53	0.91
40 gallons	0.54	0.53	0.90
50 gallons	0.53	0.50	0.88
60 gallons	0.51	0.48	0.87

The most efficient electric storage water heaters have EFs between .96 and .98. The most efficient gas storage water heaters have EFs ranging from .60 to .86. Brand names of the most efficient models are listed in *Consumer Guide to Home Energy Savings*. You can also improve the efficiency of your water heater by adding an insulating blanket, available at hardware and home improvement stores.

Water Purification

🡅 USE AN ACTIVATED CARBON (AND/OR KDF) FILTER TO REMOVE VOLATILE CHEMICALS. These include all nonparticulate substances that can be vaporized—trihalomethanes (chlorine, chloramine, chloroform, and others), chlorinated hydrocarbons, pesticides (such as DDT, dialdrin, lindane, heptachlor), radon gas, benzene, carbontetrachloride, trichloroethylene, xylene, toluene, and hundreds of others. If your water is chlorinated, you could benefit from using a carbon filter, if only to improve the taste of the water.

Activated carbon works through the process of adsorption. Each particle of carbon is like a honeycomb of minute pores that attract and trap pollutant molecules. As the water passes through and the micropores are filled, fewer and fewer pores remain vacant. As the pores become filled, the effectiveness of the carbon decreases, and when the carbon becomes fully saturated, not only do the pollutant molecules flow right through the carbon, but some of the trapped pollutants are also released into the product water. For this reason you must change the carbon regularly. Tests done by Rodale Product Testing Labs show that the effectiveness of contaminant removal declined sharply after about 75 percent of the rated life on all the filters tested. They suggest changing the carbon more often than recommended.

Activated carbon units should only be used with chlorinated municipal tap water or other treated water supplies that are free from bacteria. Activated carbon will not remove the bacteria that is present in all water, and it will grow in the unit. Some bacterial contamination can also occur even with chlorinated water, so this is another good reason to change your carbon regularly. While there is no evidence that these bacteria are a common cause of disease when they enter the body via water passed through carbon units, there is no practical way to monitor the amount or type of bacteria that may be present in home filters. Bacteriostatic filters have been embedded with silver to control bacteria growth, but EPA studies have found that the silver does not actually perform this function. And as *Consumer Reports* says, "Bacteriostasis in a carbon filter is of little usefulness. Water supplied by municipal systems doesn't contain dangerous levels of bacteria."

Carbon block is generally favored over granular carbon. Because it is compressed, more carbon fits into the same amount of space and provides more micropores for greater efficiency. In addition, there is a straining effect that removes some bacteria and heavy metals, and it has less of a tendency to breed bacteria.

KDF is a relatively new material to the home water purification field. It is a high-purity copper-zinc alloy that is melted together and flaked to make a "golden sand." It has two properties. KDF uses chemical oxidation reduction to neutralize harmful chemicals, including chlorine, hydrogen sulfide, and iron; these substances are actually chemically changed into harmless substances (chlorine to chloride, etc.), which are then mechanically filtered out. KDF also produces .04 volts of electricity and acts like little magnets to attract and permanently trap pollutants like heavy metals.

KDF is most commonly used in small shower head filters that are designed specifically to remove chlorine from shower water. Because of its excellent and long-lasting ability to neutralize chlorine, it is also used in conjunction with activated carbon, to enhance its efficiency and extend its life. Because the KDF removes most of the chlorine, the carbon in the unit is then free to take out other volatile chemicals that are not removed by the KDF. Together KDF and carbon work much more efficiently than carbon alone.

Because activated carbon and KDF units work by removing pollutants as the water passes through them, the amount of clean water available per day is limited only by how much water can flow through the tap. Activated carbon units give you lots of choices when it comes to style, design, and capacity: over-the-counter units that sit on countertops with diverter tubes to and from the tap, undersink models that dispense filtered water through either the main or an auxiliary tap, units that attach to a shower head, and large units that filter water for an entire house. One type of carbon filter I don't recommend, however, are the popular pour-through pitcher-type filters sold in many hardware, discount, and department stores. While they will remove some pollutants, *Consumer Reports* found that they removed only about half the pollutants after just twenty gallons—a practically worthless investment that will actually add pollutants to the water if the carbon pads are not changed every week or so.

Activated carbon and KDF units generally have cartridges which have a rated life of a number of gallons and an approximation of how long the cartridge will last under normal real-life circumstances. If you choose a carbon filter, it is mandatory that you replace the cartridges on a regular basis, otherwise you would be better off drinking tap water, since the carbon will unload previously removed pollutants back into the water after a certain saturation is reached. The frequency of replacement ranges from every few weeks to several months, to a year, depending on the size of the filter and the amount of pollution in your water.

Many of the most popular activated carbon units have a single cartridge that is made up of a block of carbon surrounded by a sediment prefiltering material. For both economy and efficiency, the preferred design is that which has two separate canisters—one for the sediment prefilter and the other for the carbon. Frequent change of inexpensive sediment filters can prolong the effective life of the carbon; with the single cartridge design, the entire (expensive) cartridge must be replaced.

♦ USE A DISTILLER TO REMOVE PARTICLES. Distillation has been used for years to remove microorganisms (microscopic plants and animals more commonly known as bacteria, viruses, protozoan, algae, and cysts), particulates (all the minute bits of material that do not dissolve in water: metals [such as lead], asbestos, rust, dirt, sand, and other sediments), dissolved solids (materials like fluoride, nitrates, sulfates, and salts, which decompose in water), and radioactive particles from water. Because it removes 99.9 to 100 percent of contaminants, distillation is the method preferred by scientific laboratories. While the older distillers were not designed for the removal of volatile chemicals, newer models incorporate functions that remove significant amounts of these pollutants. Used in conjunction with carbon filtration, distillation gives the highest purity of all home units—consistently, gallon after gallon.

Water distillers remove pollutants by boiling water to turn it into steam, then condensing it. Boiling the water destroys bacteria and other living organisms, leaving them behind in the boiling tank, along with particulates, dissolved solids, and radioactive particles that are too heavy to rise with the water vapor.

The major objection most people have to distillers is that they are inconvenient. Some have to be filled by hand; others have to be turned on and off manually. Regardless of the level of participation, in order to drink distilled water, you have to make it. Distillers cost money and energy to operate (about a dollar's worth of electricity to produce five gallons of water, but you never have to replace anything, and there are solar distillers) and must be cleaned regularly (depending on water composition and size of the distiller, once every two weeks to six months). They also waste water—about six to seven gallons for every gallon of distilled water they make.

Distillation is also limited in the number of gallons produced, which differ according to the size of the units. Most people who make distilled water run their distillers and then store the water in pitchers or five-gallon glass water bottles (such as the kind water used to be delivered in).

➧ REVERSE OSMOSIS. Reverse osmosis units incorporate both activated carbon and a reverse osmosis membrane. The membrane alone removes significant amounts of particulates, dissolved solids, and radioactive particles. Combined with the volatile chemical removal of carbon, a reverse osmosis unit can provide good broad-spectrum removal of most common pollutants.

There are several drawbacks, however, that make reverse osmosis systems the second best choice in terms of water purity. At best, most reverse osmosis membranes will reject pollutant molecules only to about 90 percent efficiency, and will diminish with use. The synthetic materials from which they are made can vary in consistency, and day to day efficiency might further be diminished by changes in water pressure, the pH of the water, or the temperature of the water. In addition, reverse osmosis systems are made almost entirely of plastic, and store water in a plastic-lined holding tank. The plastic does get into the water, although most is removed by the carbon filter. A reverse osmosis unit is ideal for someone who values convenience over effectiveness—it gives more than two gallons a day of somewhat pure water at the flip of a lever. It's better than drinking tap water, but is not the purest water available.

An environmental disadvantage to reverse osmosis systems (besides the toxic manufacture and ultimate disposal problems of all the plastic) is that they waste a lot of water. Only 10 to 15 percent of the water passing through the system goes through the membrane; the rest goes down the drain. About twenty gallons each day is wasted by being flushed through the unit, even if you don't use any water. Some newer types recycle the waste water, which is a good idea even if you live in an area where there is plenty of water.

Many department, discount, and hardware stores now offer drinking water filters. You may find something that is appropriate for you there; however, bear in mind that these units were probably chosen by a buyer who had little knowledge of water purification and were primarily chosen for their low price, rather than their ability to purify water.

Almost every multilevel marketer of household products sells some kind of activated carbon filter, most of which contain carbon cartridges made by the same manufacturer. These are entirely adequate carbon filters and differ more in price than performance. You can probably find similar quality for less money, however.

If you'd like some help making your decision from a knowledgeable person, look in the Yellow Pages of your telephone book under "Water Filtration and Purification Equipment." In addition to sources for filters you can install yourself, there will be listed companies that can install and maintain systems for you. You can also order water filters by mail from health and environment-oriented catalogs, or from an independent dealer such as ***Nigra Enterprises.***

Some unusually good water purification units I have recommended over the years include an enhanced carbon unit called the **Seagull IV,** the **Rain Crystal** Pyrex glass water distiller, and a three-canister reverse osmosis/carbon system with a high rejection rate called the **Water Safe WS/RO5.** A KDF shower filter can be ordered from ***Real Goods.***

Water Softeners

♠ SALT. Water softeners work by adding some type of salt to the water, which exchanges the hard calcium and magnesium ions in water for soft sodium ions, so actually any salt will do—sodium bicarbonate (baking soda) is fine. Just add enough so that the water feels "slippery." Water softeners are necessary in hard-water areas when using soap to clean and prevent soapy deposits in bathtubs and sinks and soap scum that dulls laundry.

See also CLEANING PRODUCTS.

Window Coverings

♠ NATURAL FIBER CURTAINS OR DRAPES. Natural fiber curtains and drapes are available at some curtain shops, or you can order by mail from ***Country Curtains, Crocodile Tiers,*** or ***Homespun Fabrics & Draperies*** (unbleached, undyed cotton in nice textures).

♠ WOODEN SHUTTERS. Though probably not sustainably harvested, wood is a renewable resource that is biodegradable. Purchase unfinished and finish with a natural or nontoxic PAINT.

♠ RICE PAPER SHADES. Buy these at any import store.

♠ ALUMINUM MINI-BLINDS. See if you can find some with recycled content.

Windows

♠ SALVAGE. If you need to replace your windows in a moderate climate, consider going to a salvage yard and reclaiming some windows that already exist (wood frames are much more insulative than aluminum frame windows).

♦ ENERGY-EFFICIENT. If you are replacing windows in an exceptionally cold or hot climate, the new energy-efficient (so called because they make your home more energy-efficient) windows are a wise and economical investment.

In the past, insulating glass was nothing more than two panes of glass separated by an air space, mounted in a single frame. While this doubled the insulating value to R-2 from the R-1 insulating value of a single pane of glass, new technologies do much better. Adding transparent heat-reflective coatings (called "low emissivity" or "low-e" coatings) increases the insulating value to R-3. Blowing colorless, inert argon gas between the panes of sealed low-e windows brings them up to R-4. A heat-reflective coating on clear, colorless film mounted inside a double-pane window makes it R-4.5. The new Superglass system has an insulating value of R-8. The next frontier for improving the R-value of windows will be to add insulation at the edge of the window. In addition to improving energy efficiency, new windows can keep your house cooler, block ultraviolet rays, and retard sound transmission.

When shopping for windows, compare R-values carefully, as some manufacturers list just the R-value through the center of the glass, while others provide unit R-values as well, which take into account heat loss through the edge of the windows.

Insulated shades or window blankets make your windows even more energy-efficient than a single layer of fabric. Usually these are made from synthetic materials, but you can order cotton window blankets from ***Crocodile Tiers.***

Wine

♦ ORGANICALLY GROWN GRAPES. Many natural food stores now sell some very nice wines made from organically grown grapes. Some are imported from France, Germany, and Italy, but there are quite a few made in California.

A handful of small producers have made organic wines since the early 1980s, but now larger companies—including Gallo, Fetzer, Sutter Home, and Buena Vista—are also using organic grapes. More than sixty California wineries (10 percent of the total) now have vineyards certified by California Certified Organic Farmers and many others are using organic practices but choose not to be certified for a variety of reasons. Wineries in other states are also going organic.

While small wineries have been carefully labeling and advertising their special growing practices, wines made by the larger vintners may not have any indicators on the label (don't assume though, that all wines made by these companies are organic).

One major problem with labeling wines made from organically grown grapes is the presence of sulfites in the wine. Sulfites can cause allergic reactions in some people, and are generally considered a food additive to be avoided. Virtually all wines contain some sulfites, because sulfite is naturally formed in the process of fermentation. However, some winemakers also add sulfites to inhibit oxidation and spoilage (a practice that has been used for centuries). If a wine contains more than 10 parts per million (ppm) sulfite, the statement "contains sulfites" must appear on the label. Most wines contain 30 to 150 ppm.

But in America, use of sulfites in organic wine is forbidden by law. A few wines are sulfite-free, but most vintners think sulfites are necessary to make a good wine. So a proper label could say "made with organically grown grapes" and "contains sulfites," though many large companies will probably just keep their old labels, even as they switch to organically grown grapes. Many winemakers just see going organic as the responsible thing to do, but are not certified so as to maintain flexibility and the ability to use pesticides occasionally as they choose.

The Organic Grapes into Wine Alliance has its own standards for organic wine, which include being made from certified organic grapes, having low sulfite levels, using no toxic materials in packaging (including no tin-lead capsules), and using recycled glass bottles.

🡅 NATURAL. Virtually all wines made from nonorganic grapes, both domestic and imported, contain sulfites. These can cause fatal anaphylactic allergic reactions in those who are sensitive to them, but for the general public, their danger is no more harmful than the pesticide residues that are present in nonorganic wines.

See also FOOD.

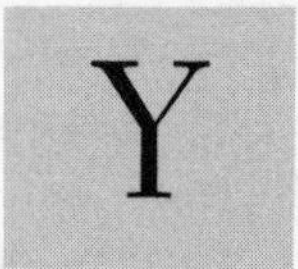

Yarn

🡅 NATURAL FIBERS. Look for yarns made from cotton, silk, wool, and other natural fibers in your local yarn shop.

🡅 NATURAL DYES. Yarns colored with natural dyes are becoming more and more popular. Ask around locally for someone who is a spinner or weaver, who might also dye his yarns with plants. ***Sajama Alpaca*** imports alpaca yarns handspun and dyed with plants indigenous to Bolivia. The wool comes from animals who live in their own habitat on the open range, kept by people who spin the yarn themselves. **FoxFibre Colorganics**' naturally brown and green yarns can be ordered by mail from ***Vreseis.***

Yogurt

🡅 ORGANIC. If it is available in your area, you can find yogurt made from organically or biodynamically produced milk at your natural food store.

🡅 NATURAL. Plain yogurt is additive-free, but the supermarket brands of flavored yogurt generally contain sugar or an artificial sweetener. Buy plain yogurt and sweeten it yourself with fruit or honey, or buy flavored yogurts at your natural food store.

See also FOOD.

Recommended Reading

Magazines and Newsletters

The publications listed here will lead you to information on more sustainable products and life-style choices. However, I can't honestly say that each of these publications will always agree with the philosophy outlined in this book. Use them for leads and ideas, but make your own decisions. I also suggest that you subscribe to a regional life-style magazine—here in California we have one called Sunset *that covers gardening, travel, cooking, do-it-yourself home projects, etc.—for these can tell you a lot about what is going on locally and is appropriate for your locale.*

American Harvest (3 Gold Center, Suite 221, Hoffman Estates, IL 60195, 708-934-7655). A newsletter focused on "celebrating the fresh fruits and vegetables harvested from America's gardens, big and small." For gardeners and cooks, with a "green" philosophy.

Building With Nature (P.O. Box 369, Gualala, CA 95445, 707-884-4513). A bimonthly newsletter on sustainable architecture. Geared for design professionals, but easy to read and interesting to anyone who is interested in the practicalities of natural building.

Buzzworm's Earth Journal (P.O. Box 6853, Syracuse, NY 13217, 800-825-0061). Bimonthly magazine that "strives to offer balanced and comprehensive coverage" on national and international environmental issues, including green living.

Catalyst (P.O. Box 1308, Montpelier, VT 05601, 802-228-7948). One of the pioneering publications on socially responsible investing, with a sustainable view.

Consumer Reports (Consumers Union, 101 Thurman Avenue, Yonkers, NY 10703-1057, 800-234-1645). A monthly magazine that provides consumers information and advice on goods, services, health, and personal finances. While historically their product testing and ratings have not had an environmental slant, in more recent issues thay have turned their objective eye to testing some products for environmental claims. (They're even starting to print their sixty million copies a year on recycled paper.)

Country Life (The Creamery, Charlotte, VT 05445, 800-344-3350). Bimonthly magazine geared to people who live close to the earth in the country. Topics are more focused on gardening and practical householding.

Essential Living (Maia Institute, R.R. 1, Box 1310, Moretown, VT 05660, 802-244-1309). A thoughtful, small bimonthly newsletter about "examining our personal lives with an eye towards making changes that will both decrease the stresses we put on our planet and increase the joy and satisfaction in our own lives."

E—The Environmental Magazine (P.O. Box 6667, Syracuse, NY 13217, 800-825-0061). A bimonthly magazine that is a clearinghouse of news, information, and commentary on environmental issues to inform and inspire individuals to bring about improvements. Not many articles on products or life-style, but has environmental product ads.

Garbage (P.O. Box 56519, Boulder, CO 80322-6519, 800-274-9909). A bimonthly magazine that calls itself "the practical journal for the environment." It has a real focus on garbage, but also has a fair number of articles and a lot of advertising for consumer products.

Green Alternatives (P.O. Box 28, Annandale-on-Hudson, NY 12504, 914-876-6525). A bimonthly how-to, hands-on green living, consumer-oriented magazine. Edited by Annie Berthold-Bond, author of *Clean & Green.*

Home Energy (2124 Kittredge Street #95, Berkeley, CA 94794, 510-524-5405). Bimonthly magazine on residential energy conservation, including new energy-saving technologies and tips for home energy savings.

Home Power (P.O. Box 130, Hornbrook, CA 95044-0130, 916-475-3179). Bimonthly magazine about residential renewable energy systems.

In Business (P.O. Box 323, Emmaus, PA 18049, 215-967-4135). A bimonthly journal oriented to small manufacturers and retailers of environmental products.

In Context (Context Institute, P.O. Box 11470, Bainbridge Island, WA 98110, 206-842-0216). "A quarterly of humane sustainable culture" envisions, explores, and clarifies the many ways cultures can be both humane and sustainable—and how we can get there. If you want one publication about our emerging sustainable society, this is it.

Interior Concerns (P.O. Box 2386, Mill Valley, CA 94942, 415-389-8049). A bimonthly newsletter providing information for designers, architects, builders, and homeowners on environmental products, issues, and industry changes. They also have an *Interior Concerns Resource Guide* that lists products, manufacturers, consultants, and other resources.

Mother Earth News (24 E. 23rd Street, New York, NY 10010, 800-937-4287). "The original country magazine" was one of the early pioneer publications on sustainable living. Geared for people who live in the country, but useful for anyone wanting to live more sustainably.

Mothering (P.O. Box 1690, Santa Fe, NM 87504, 505-984-8116). A comprehensive, quarterly magazine on natural mothering. Includes articles written by mothers and resources for all types of natural products used by babies, children, and mothers.

Natural Health (P.O. Box 57329, Boulder, CO 80322-7320, 800-666-8576). The bimonthly "guide to well-being" focuses mainly on natural healing, with natural living a part of overall good health practices.

NewAge Journal (P.O. Box 52375, Boulder, CO 80321-3275, 800-234-4556). Bimonthly magazine that covers everything about the New Age, including natural living along with articles on spirituality and personal growth.

Organic Gardening (Box 7320, Red Oak, IA 51591-0320, 800-666-2206). The classic bimonthly organic gardening magazine since 1942.

Reuse News (c/o Sherlock Enterprises, 609 Hobart Road, Hanover, PA 17331). Monthly newsletter offers tips on reusing everything from chopsticks to panty hose.

Utne Reader (P.O. Box 1974, Marion, OH 43305, 800-736-UTNE, 614-382-3322). "The best of the alternative press," this bimonthly magazine addresses many environmental and social issues by reprinting articles from other publications.

Vegetarian Times (P.O. Box 570, Oak Park, IL 60303, 800-435-9610). Monthly magazine that promotes a natural, vegetarian life-style.

Whole Earth Review (27 Gate Five Road, Sausalito, CA 94965, 415-332-1716). This quarterly magazine, which emerged from the original *Whole Earth Catalog* twenty-five years ago, is still on the leading edge. Includes essays and reviews about tools and ideas relevant to creating a sustainable world "in the new millennium."

WorldWatch (WorldWatch Institute, 1776 Massachusetts Avenue NW, Washington, DC 20036, 202-452-1999). Tracks key indicators of the earth's well-being, monitoring evolving environmental trends and making connections between the world's economic system and environmental systems.

Yoga Journal (2054 University Avenue, Berkeley, CA 94704, 415-841-9200). Monthly magazine that includes natural living as part of daily yoga practice.

Books

Agenda 21 by the United Nations Environment Programme. This is a United Nations document of many volumes, but there is a summary. Includes twenty-seven principles of sustainable development.

Artist Beware: The Hazards and Precautions in Working with Art and Craft Materials by Michael McCann (New York: Watson-Guptill Publications, 1979). Provides information on toxic substances found in art materials, and how you can set up a nontoxic studio.

Beyond the Limits: Confronting Global Collapse, Envisioning a Sustainable Future by Donella Meadows, Dennis Meadows, and Jørgen Randers (Post Mills, VT: Chelsea Green Publishing, 1992). The sequel to the international best-seller of twenty years ago, *The Limits to Growth.* Gives computer-generated information on how our way of life is unsustainable.

Caring for the Earth: A Strategy for Sustainable Living by The World Conservation Union, United Nations Environment Programme, and the World Wildlife Fund for Nature (Switzerland, 1991). Concrete suggestions for sustainable changes.

Catalog of Healthy Food, The by John Tepper Marlin, Ph.D. (New York: Bantam Books, 1990). A comprehensive guide to healthy food and food hazards, including listings of organic farms, products, and vendors.

Children's Art Hazards by Lauren Jacobsen (Natural Resources Defense Council, 40 West 20th Street, New York, NY 10011, 212-727-2700). A basic overview of children's art materials and health hazards, and art precautions for children under twelve years of age.

Clean & Green: The Complete Guide to Nontoxic and Environmentally-Safe Household Cleaning by Annie Berthold-Bond (Woodstock, NY: Ceres Press, 1990). More than five hundred recipes for natural, homemade cleaning products.

Clearcutting: A Crime Against Nature by Edward C. Fritz (Austin, TX: Eakin Press, 1989). Convincing evidence that shows how bad clear-cuts are and examples of successful, sustainably harvested forests. Includes list of sustainable forests.

Clinical Toxicology of Consumer Products, The by R. E. Gosselin et al. (Baltimore: The Williams & Wilkins Company, 1984). A big, expensive book that is in the reference section of most libraries. Lists many brand-name products and their toxic ingredients that do not appear on the labels, as well as much toxicological data as to their health effects.

Closing Circle, The by Barry Commoner (New York: Bantam Books, 1974). Dated, but still one of the best books about how our current way of life is not sustainable.

Common Sense Pest Control: Least-toxic Solutions for Your Home, Garden, Pets, and

Community by William Olkowski, Sheila Daar, and Helga Olkowski (Newton, CT: Taunton Press, 1991). This is the definitive book on nontoxic pest controls—seven hundred pages of information from the most knowledgeable people in the field. Compiles their years of experience at the Bio-Integral Resource Center.

Consumer Dictionary of Cosmetic Ingredients, A by Ruth Winter (New York: Crown Publishers, 1984). Alphabetical listing of thousands of cosmetic ingredients, how they're used, and their health effects.

Consumer Dictionary of Food Additives, A by Ruth Winter (New York: Crown Publishers, 1989). Alphabetical listing of thousands of food additives, how they're used, and their health effects.

Consumer Guide to Home Energy Savings by Alex Wilson and John Morrill. Published by American Council for an Energy-Efficient Economy (1001 Connecticut Avenue NW, Suite 535, Washington, DC 20036, 202-429-8873) in cooperation with *Home Energy* magazine. The most practical and easy-to-read guide to home energy savings I've found; includes extensive brand-name listings of the most energy-efficient appliances available. ACEEE also has a whole catalog of other books about energy policy and conservation (though some are quite technical).

Drinking Water Hazards: How to Know If There Are Toxic Chemicals in Your Water and What to Do If There Are by John Cary Stewart (Hiram, OH: Envirographics, 1990). A comprehensive and well-documented guide to water pollutants, water testing and treatment, and bottled water.

Ecology of Commerce: A Declaration of Sustainability, The by Paul Hawken (New York: HarperCollins, 1993). Outlines how business and economic structures need to change to become more sustainable.

Economics as If the Earth Really Mattered by Susan Meeker-Lowry. (Philadelphia: New Society Publishers, 1988). The classic book on socially responsible investing.

Energy Saver's Handbook for Town and City People, The by the Scientific Staff of the Massachusetts Audubon Society (Emmaus, PA: Rodale Press, 1982). Though its chapters on renewables are dated, this book gives good, in-depth, easy-to-understand background information on all energy-saving options.

Energy Unbound: A Fable for America's Future by L. Hunter Lovins, Amory Lovins, and Seth Zuckerman (San Francisco: Sierra Club Books, 1986). An easy-to-read story about a housewife appointed national Secretary of Energy, which clearly explains how the United States can maintain energy abundance and lower energy costs.

Evaluation of Environmental Marketing Terms in the United States (EPA document #741-R-92-003, 1993. Available from the EPA Office of Pollution Prevention and Toxics, Washington, DC 20460). Gives an overview of consumer awareness and understanding of environmental marketing terms, and summarizes the varying views regarding the definitions of the most common environmental buzzwords.

Forest Journey: The Role of Wood in the Development of Civilization, A by John Perlin (Cambridge, MA: Harvard University Press, 1991). A fascinating account of how different societies have used wood. Tells just where our forests have gone.

Global Ecology Handbook: What You Can Do About the Environmental Crisis, A Guide to Sustaining the Earth's Future, The by the Global Tomorrow Coalition (Boston: Beacon Press, 1990). Lots of details on worldwide environmental problems, some sustainable solutions.

Green Business: Hope or Hoax?: Toward an Authentic Strategy for Restoring the Earth edited by Christopher and Judith Plant (Philadelphia: New Society Publishers, 1991). Presents many opinions on the value of green marketing and what really needs to be done to improve the environment.

Green Groceries: A Mail-Order Guide to Organic Foods by Jeanne Heifetz (New York: HarperCollins, 1992). Contains more than 275 mail-order sources for organically grown foods.

Green PC, The by Steven Anzovin (Blue Ridge Summit, PA: Windcrest/McGraw-Hill, 1993). One hundred practical things you can do to reduce the environmental hazards of computing.

Greenpeace Guide to Paper, The (order from Greenpeace, 1436 U Street NW, Washington, DC 20009, 202-462-1177). Everything about the environmental effects of paper.

Home Economics by Wendell Berry. This book is out of print, but look for it in the library or in used book stores. This, and all books by the author, particularly also *The Unsettling of America* (San Francisco: Sierra Club Books, 1986) give profound insight into the connections between our daily lives and the ways of nature.

How Much Is Enough? by Alan Durning (New York: Norton, 1992). An examination of our overconsumptive society.

Lessons from Nature: Learning to Live Sustainably on the Earth by Daniel D. Chiras (Washington, DC: Island Press, 1992). One man's vision of a sustainable society. Excellent and easy-to-read introductory book to sustainable living.

New Woodburner's Handbook: A Guide to Safe, Healthy & Efficient Woodburning, The by Stephen Bushway (Pownal, VT: Storey Communications, 1992). All the details about heating with wood and woodburning apparatus.

Nontoxic, Natural, & Earthwise by Debra Lynn Dadd (Los Angeles: Jeremy P. Tarcher, 1990). A veritable Yellow Pages of environmental products, with do-it-yourself formulas, brand names, and mail-order resources.

Nontoxic Home & Office, The by Debra Lynn Dadd (Los Angeles: Jeremy P. Tarcher, 1992). A primer on toxics in the home and office, and nontoxic alternatives. A good book to begin with to learn about toxics.

Our Common Future by the World Commission on Environment and Development (New York: Oxford University Press, 1987). The book that made sustainable development a global issue. Outlines issues and makes recommendations.

Pesticide Alert: A Guide to Pesticides in Fruits and Vegetables by Lawrie Mott and Karen Snyder of the Natural Resources Defense Council (San Francisco: Sierra Club Books, 1987). Pesticides commonly used on produce and their health effects.

Rainforest Book: How You Can Save the World's Rainforests, The by Scott Lewis with the Natural Resources Defense Council (Los Angeles: Living Planet Press, 1990). A concise and easy-to-read guide to what you can do for the rain forests.

Recycled Papers: The Essential Guide by Claudia G. Thompson (Cambridge, MA: MIT Press, 1992). A beautiful book with many photographs and illustrations that gives much technical data on the paper-making and recycling process, as well as environmental issues. Primarily written for design professionals, it also includes chapters on choosing and specifying recycled papers, and design.

Recycler's Handbook: Simple Things You Can Do, The by The Earthworks Group (Berkeley, CA: EarthWorks Press, 1990). A simple but comprehensive overview of recycling options.

Shopper's Guide to Recycled Products (order from Californians Against Waste Foundation, 909 12th Street, Suite 201, Sacramento, CA 95814, 916-443-8317). Lists a variety of national brand-name products that have recycled content or come in recycled packaging. *Guide to Recycled Printing and Office Paper* lists more than two hundred paper products with post-consumer content.

Shopping for a Better World: The Quick and Easy Guide to Socially Responsible Supermarket Shopping by Alice Tepper Marlin et al., The Council on Economic Priorities

(New York: Ballantine Books, 1991). Over a thousand supermarket products rated for social responsibility.

Soul of a Business: Managing for Profit and the Common Good, The by Tom Chappell (New York: Bantam Books, 1993). How Tom's of Maine is a socially responsible company.

State of the World: A Worldwatch Institute Report on Progress Toward a Sustainable Society by Lester R. Brown et al. (New York: W. W. Norton, 1991). A collection of essays on major environmental problems and emerging sustainable solutions.

Status Report on the Use of Environmental Labels Worldwide (EPA document #742-R-93-003, 1993. Available from the EPA Office of Pollution Prevention and Toxics, Washington, DC 20460). A review of official and unofficial environmental labeling programs and their product criteria.

Sustainable Harvest and Marketing of Rain Forest Products by Mark Plotkin and Lisa Famolare of Conservation International (Washington, DC: Island Press, 1992). A technical but comprehensive evaluation of the potential value of rain-forest products.

Taking Out the Trash: A No-Nonsense Guide to Recycling by Jennifer Carless (Washington, DC: Island Press, 1992). A complete review of issues regarding recycling, the recycling industry, regulations, and market.

Toxics A to Z: A Guide to Everyday Pollution Hazards by John Harte et al. (Berkeley, CA: University of California Press, 1991). One of the most complete books I've seen that explain the problems of toxicity in everyday language. Includes detailed health and environmental toxicology information for over one hundred commonly encountered toxics.

Use of Life Cycle Assessment in Environmental Labeling, The (EPA document #742-R-93-003, 1993. Available from the EPA Office of Pollution Prevention and Toxics, Washington, DC 20460). Documentation of the methodologies used by independent, third-party environmental labeling programs for the development of criteria for certification and the use of life cycle assessment methodologies being used in environmental labeling programs worldwide.

We Build the Road as We Travel by Roy Morrison (Philadelphia: New Society Publishers, 1991). A detailed account of the history and workings of the Mondragon cooperatives in Spain.

Wildwood: A Forest for the Future by Ruth Loomis with Merv Wilkinson (Reflections Publisher, P.O. Box 178, Gabriola, BC V0R 1X0). A detailed description of a sustainably managed forest.

World Resources by the World Resources Institute (Baltimore, MD: World Resources Institute Publications). Published yearly, gives information on the status of all the renewable and nonrenewable resources around the world. If not available in your local bookstore, call 1-800-822-0504.

Directory of Greenstores

Greenstores—as they call themselves—are small shops specializing in environmentally aware goods. While the selection of products differs from store to store, most carry a wide array of products that are better for the environment than their counterparts found in regular stores.

A note of caution, however. In a greenstore, as anywhere else, the buyer must beware. Most of these stores are labors of love financed by the life savings of well-meaning individuals. More likely than not, you'll find the newest and the most sustainable products in these stores first. But don't assume that the environmental standards of the shop owner meet your (or my) standards. Read the labels, and evaluate the environmental integrity of each product for yourself.

ALABAMA
Earth Friendly of Huntsville
2357 Whitesburg
Huntsville, AL 35801
205-536-5602

ALASKA
Alaska Green Goods
535 Second Avenue, Suite 103C
Fairbanks, AK 99701
907-452-4426

ARIZONA
Ecology Store
3615 N. Campbell Avenue
Tucson, AZ 85719
602-327-3235

CALIFORNIA
Earth Options
6930 McKinley
Sebastopol, CA 95472
800-269-1300

Earthsake
1805 4th Street
Berkeley, CA 94710
510-848-0484

Ecology Center
2530 San Pablo Avenue
Berkeley, CA 94702
510-548-2220

Enviro Care
26544 Carmel Rancho Shopping Center
Carmel, CA 93923
408-373-8420

EnvironGentle
246 N. Highway 101
Encinitas, CA 92024
619-753-7420

Green Store
2232 Sunset Cliffs Boulevard
San Diego, CA 92107
619-225-1083

Green World Mercantile
2340 Polk Street
San Francisco, CA 94109

Greener Alternatives
914 Mission Street, Suite A
Santa Cruz, CA 92107
408-423-7420
415-771-5717

Real Goods
966 Mazzoni
Ukiah, CA 95482
707-468-9430

Solutions
928 9th Street
Arcata, CA 95521
707-822-6972

COLORADO
Alfalfa's
141 E. Meadow Drive
Vail, CO 81657
303-476-1199

Daily Planet
332 N. Tejon
Colorado Springs, CO 80903
719-633-7163

Ecology House
1441 Pearl Street
Boulder, CO 80302
303-939-0204

Partners for a Better Earth
423 W. Alcorn Avenue
Estes Park, CO 80517
303-586-0800

Planetary Solutions
3158 28th Street
Boulder, CO 80301
303-442-6228

CONNECTICUT
Earth Animal
606 Post Road East
Westport, CT 06880
203-222-7173

FLORIDA
Blue Moon Trader
Route 5, P.O. Box 129
Big Pine Key, FL 33043
305-872-8864

Earth Sensitive Products
2910 Kerry Forest Parkway, Suite D2
Tallahassee, FL 32308
904-893-2032

Eco Store
2421 Edgewater
College Park, FL 32804
407-426-9949

Environeers
5341 Fruit Ville Road
Sarasota, FL 34232
813-371-4492

GEORGIA
Common Pond
1402 N. Highland Avenue #3
Atlanta, GA 30306
404-876-6368

HAWAII
Planet Saviors
444 Hana Highway
Kahalui, Maui, HI 96732
808-871-1561

Solarworks
Oceanview Town Center
Naalehu, HI 96777
808-929-9820

ILLINOIS
Alternatives
1306 E. Seiberling
Peoria Heights, IL 61614
309-685-2866

Alternatives
114 Lincoln Square
Urbana, IL 61801
217-328-2256

INDIANA
EcoShop
5884 E. 82nd Street
Indianapolis, IN 46250
317-849-6753

KANSAS
Natural Habitat Healthy Home Store
141 N. Rock Island
Wichita, KS 67202
316-265-4785

Simple Goods General Store
735 Massachusetts Avenue
Lawrence, KS 66044
913-841-8321

LOUISIANA
Eco-Habits
7913 Maple Street
New Orleans, LA 70118
504-866-8870

MAINE
Ecology House
Maine Mall
South Portland, ME 04106
207-775-7441

Ecology House
49 Exchange Street
Portland, ME 04101
207-775-4871

Maine Concern
Route 17 and 90 RR 1, P. O. Box 975
West Rockport, ME 04865
207-236-8628

Natural Living Center
570 Stillwater Avenue
Bangor, ME 04401
207-990-2646

Natural Living Center
471 Wilson Street
Brewer, ME 04412
207-989-7996

MARYLAND
Environmentally Sound Products
8845 Orchard Tree Lane
Towson, MD 21286
800-886-5432

Green Goods
7001 Carol Avenue
Takoma Park, MD 20912
301-891-1111

MASSACHUSETTS
Annie's Garden Store
515 Sunderland Road
Amherst, MA 01002
413-549-6359

Earthwise Trading Company
10 Pleasant Street
Newburyport, MA 01950
508-462-6922

Good Seed
138 Central Avenue
Seekonk, MA 02771
508-399-7333

Green Living
317 Chestnut Street
Needham, MA 02192
617-449-7700

Green Planet
22 Lincoln Street
Newton, MA 02161
617-332-7841

Options
17 Strong Avenue
North Hampton, MA 01060
413-584-0010

MINNESOTA

Restore the Earth Store
2204 Hennepin Avenue S.
Minneapolis, MN 55405
612-374-3738

Terra Firma
809 W. 50th Street
Minneapolis, MN 55419
612-824-5673

NEVADA

Earthgoods
276 Kingsbury Grade
Stateline, NV 89449
702-588-8006

NEW HAMSPHIRE

Ecology House
99 Rockingham Boulevard
Salem, NH 03079
603-893-7712

Hanover Consumer Co-op
45 S. Park Street
Hanover, NH 03755
603-643-2667

Sunweaver
RR 2, P.O. Box 155
Northwood, NH 03261
603-942-5863

NEW JERSEY

Green Tree
664 Washington Street
Cape May, NJ 08204
609-884-0065

Whole Earth Center
360 Nassau Street
Princeton, NJ 08540
609-924-7377

NEW YORK

Boondocks
490 Piermont Avenue
Piermont, NY 10968
914-365-2221

Dreams East
1 Tower Street
Roslyn, NY 11576
516-484-5384

Earth Doctor
889A New Loudon Road
Latham, NY 12110
518-782-1648

Earth General
72 Seventh Avenue
Brooklyn, NY 11217
718-398-4648

Earth Shop
800 New Louden Road
Latham, NY 12110
518-783-3163

Earth Smarts
126 Mamaroneck Avenue
Mamaroneck, NY 10543
914-698-6969

Energy Clearinghouse
501 W. Fayette Street
Syracuse, NY 13204-2929
315-470-3320

Environmentally Yours
215 Katonah Avenue
Katonah, NY 10536
914-232-0382

Environmental Options
601 New Loudon Road
Latham, NY 12110
518-783-3163

Good Company
715 Monroe Avenue
Rochester, NY 14607
716-244-5719

Nature Nosh
119 Canal Street
Ellenville, NY 12428
914-647-5442

Ozone Brothers
1 Sumner Park
Rochester, NY 14607
716-244-7830

Pre-cycled
Lakeview Plaza, Route 22
Brewster, NY 10509
914-278-7611

Terra Verde Trading Company
72 Spring Street
New York, NY 10012
212-925-4533

Warm Store
12 Tannery Brook Road
Woodstock, NY 12498
914-679-4242

NORTH CAROLINA

Earthwares
Carr Mill Mall, 101 E. Weaver Street
Carrboro, NC 27510
919-929-7844

For Nature's Sake
129 Turner Street
Beaufort, NC 28516
919-728-5051

Full Circle Paper
2830 Hillsborough Road
Durham, NC 27705
919-286-0140

Salt Marsh
Kitty Hawk Connection
Nags Head, NC 27959
919-480-1116

OHIO

Better Earth
4705 N. High Street
Columbus, OH 43214
614-261-8866

Earth Naturally
19126 Old Detroit Road
Rocky River, OH 44116
216-333-2705

Eco-Operatives
4070 Woodman Drive
Kettering, OH 45440
513-294-1111

Green Earth Store
235 Xenia Avenue
Yellow Springs, OH 45387
513-767-9349

OREGON
Cabin Fever
Sunriver Village Building #25D
Sunriver, OR 97705
503-593-5675

Down to Earth
500 Olive Street
Eugene, OR 97401
503-485-5932

Earth Mercantile
6345 SW Capitol Highway
Portland, OR 97201
503-246-4935

Greater Goods
515 High Street
Eugene, OR 97401
503-485-4224

PENNSYLVANIA
Harrisburg Environmental Center
3972 Jonestown Road
Harrisburg, PA 17109
717-545-0204

Kind Earth
Doylestown Shopping Center, Route 611
Doylestown, PA 18901
215-345-1633

911 Earth
243 Desmond Street
Sayre, PA 18840
717-888-3297

RHODE ISLAND
World Store
16 W. Main Street
Wickford, RI 02852
401-295-0081

SOUTH DAKOTA
Down to Earth
609 Mt. Rushmore Road
Rapid City, SD 57701
605-342-4883

TENNESSEE
Only One Earth
340 Frasier Avenue
Chattanooga, TN 37405
615-756-3466

TEXAS
Basic Needs
5360 W. Vickery
Fort Worth, TX 76107
817-377-1947

Ecowise
1714A South Congress
Austin, TX 78704
512-326-4474

VERMONT
Earth Advocate
Route 7A, RR2, P.O. Box 2432
Sunderland, VT 05250
802-362-2766

EcoSentials
69 Elliot Street
Brattleboro, VT 05301
802-257-9377

Seventh Generation Store
176 Battery Street
Burlington, VT 05401
802-658-7770

VIRGINIA
Down to Earth
Main Street
Sperryville, VA 22740
703-948-6257

Tomorrow's World
5978 Virginia Beach Boulevard
Norfolk, VA 23502
804-461-4739

WASHINGTON
Eco-Logical Wisdom
1705 N. 45th Street
Seattle, WA 98103
206-548-1334

Eco Origins
1530 First Avenue
Seattle, WA 98101
206-467-7745

EnviResource
110 Madison Avenue
Bainbridge Island, WA 98110
206-842-9785

Directory of Mail-Order Catalogs

Abundant Life Seed Foundation (P.O. Box 772, Port Townsend, WA 98368, 206-385-5660, $2.00 donation for catalog). Open-pollinated and organic seeds.

Acres U.S.A. (P.O. Box 9547, Kansas City, MO 64133, 816-737-0064, free catalog). Books on organic gardening and farming.

After the Stork (1501 12th Street NW, Albuquerque, NM 87104, 800-333-5437, 505-243-9100, free catalog). Natural clothing and other products for babies and children.

agAccess (P.O. Box 2008, Davis, CA 95617, 916-756-7177, free catalog). Books on organic gardening and farming.

Aireox Research Corporation (P.O. Box 8523, Riverside, CA 92515, 909-689-2781, free catalog). Air purification devices.

Alexandra Avery (68183 Northrup Creek Road, Birkenfeld, OR 97016, 503-236-5926, free catalog). Organically grown cosmetic products.

AllerMed Corporation (31 Steel Road, Wylie, TX 75098, 214-442-4898, free catalog). Air purification devices.

Atlantic Recycled Paper Company (P.O. Box 39179, Baltimore, MD 21212, 800-323-2811, 410-323-2676, free catalog). Specializes in recycled paper products with high post-consumer content.

Baby Bunz & Company (P.O. Box 1717, Sebastopol, CA 95473, 707-829-5347, free catalog). Natural baby products.

Backwoods Solar Electric Systems (8530 Rapid Lightning Creek Road, Sandpoint, ID 83864, 208-263-4290, free catalog). Renewable energy systems.

Baubiologie Hardware (P.O. Box 3217, Prescott, AZ 86302, 602-445-8225, free catalog). Air purification devices and general hardware products.

Berea College Student Craft Industries (P.O. Box 2347, Berea, KY 40404, 800-347-3892, catalog $2.00). Traditional local Appalachian crafts made for fund-raising by rural Appalachian college students.

Biobottoms (P.O. Box 6009, Petaluma, CA 94953, 707-778-7945, free catalog). Natural products for babies and children.

Brody Enterprises (9 Arlington Place, Fair Lawn, NJ 07410, 800-GLU-TRAP, free catalog). Nontoxic household pest controls.

Brush Dance (100 Ebbtide Avenue, Building No. 1, Sausalito, CA 94965, 800-531-7445, 415-331-9030, free catalog). Beautiful greeting cards and wrapping papers "that renew our spirit and environment." Printed on recycled paper with soy-based ink.

Bug Off (Route 2, P.O. Box 248C, Lexington, VA 24450, 703-463-1760, free catalog). **Naturally Free** herbal repellents for people and pets.

Cal Ben Soap Company (9828 Pearmain Street, Oakland, CA 94108, 415-638-7091, free catalog). Natural soap-based products for cleaning and personal care.

Chambers (100 North Point Street, San Francisco, CA 94133, 800-334-9790, free catalog). Luxurious natural fiber bed and bath linens.

Clear Light (P.O. Box 551, State Road 165, Placitas, NM 87043, 505-867-2381, free catalog). Sustainably harvested cedar products.

Clothcrafters (P.O. Box 176, Elkhart Lake, WI 53020, 414-876-2112, free catalog). Inexpensive natural fiber household products.

Coast Filtration (142 Viking Avenue, Brea, CA 92621, 714-990-4602, free catalog). Water purification devices.

Company Store, The (500 Company Store Road, La Crosse, WI 54601, 800-356-9367, 608-785-1400, free catalog). Reasonably priced natural fiber pillows and comforters in many styles.

Coop America (2100 M Street NW, Suite 310, Washington, DC 20063, 800-424-2667, 202-872-5307, free catalog). A whole variety of nice things guaranteed to "come from groups which are socially and environmentally responsible and which practice a spirit of cooperation in the workplace."

Corns (Route 1, P.O. Box 32, Turpin, OK 73950, 405-778-3615, catalog $1.00). Organic and open-pollenated seeds.

Country Curtains (At the Red Lion Inn, Stockbridge, MA 01262, 800-937-1237, free catalog). Cotton curtains in many styles.

Crocadile Tiers (402 N. 99th Street, Mesa, AZ 85207, 602-380-3416, free catalog). Cotton window blankets and curtains, and other natural fiber products. Caters to chemically sensitive people.

Deep Diversity (P.O. Box 190, Gila, NM 88038 free catalog). Organically grown and wildcrafted open-pollinated seeds.

Deer Valley Farm (RD 1, P.O. Box 173, Guilford, NY 13780, 607-764-8556, catalog 75¢). Organically grown food.

Dry Creek Herb Farm (13935 Dry Creek Road, Auburn, CA 95603, 916-878-2441, free catalog). Organically grown and wildcrafted herbal body care products.

Dutch Mill Cheese Shop (2001 N. State Road 1, Cambridge City, IN 47327, 317-478-5847, free catalog). Cheese made by the Amish.

E. L. Foust Company (P.O. Box 105, Elmhurst, IL 60126, 800-225-9549, free catalog). Air purification devices.

Earth Care (Ukiah, CA 95842-8507, 800-347-0070). General merchandise with an emphasis on recyled paper.

Ecollection (P.O. Box 182268, Chattanooga, TN 37422–7268, 800-423-6335, free catalog). "Socially and environmentally responsible" clothing for women.

Ecology Action (18001 Schafer Ranch Road, Willits, CA 95490, 707-459-6410, free catalog). Organic gardening supplies and seeds.

Ecos Paint Company (P.O. Box 375, Saint Johnsbury, VT 05819, 802-748-9144, free catalog). Importer of Ecos VOC-free paint.

Electro Automotive (P.O. Box 1113, Felton, CA 95018–1113, 408-429-1989, free catalog). Supplier of parts and information for building your own electric car.

Emel Organic Clothing (7913 Maple Street, New Orleans, LA 70118, 504-866-8870, free catalog). Mainstream designs for women made of organic cotton.

Erwin's Bee Farm (33618 Jenkins Road, Cottage Grove, OR 97424, 503-942-7061, catalog $1.00). Beeswax candles in a variety of sizes, shapes, and colors.

Everything Under the Sun (P.O. Box 663, Winters, CA 95694, 916-795-5256, free catalog). Sun-dried, organically grown dried fruits.

Fiber Naturals (710 Daniel Shays Highway, Route 202, New Salem, MA 01355, 800-342-3707, catalog $3.00 [refundable]). Unbleached, undyed, and organic natural fiber fabrics.

Garden City Seeds (1324 Red Crow Road, Victor, MT 59875–9713, 406-961-4837, catalog $2.00). Organically grown and open-pollinated vegetable seeds suitable for northern climates.

Gardener's Supply (128 Intervale Road, Burlington, VT 15401, 800-876-5520, 802-863-1700, free catalog). Organic gardening supplies.

Gardens Alive! (5100 Schenley Place, Lawrenceburg, IN 47025, 812-537-8650, free catalog). Organic gardening supplies.

Garnet Hill (262 Main Street, Franconia, NH 03580, 800-622-6216, 603-823-5545, free catalog). High-quality natural fiber bedding and clothing.

Gold Mine Natural Food Company (1947 30th Street, San Diego, CA 92102, 800-475-FOOD, 619-234-9711, free catalog). A complete catalog of organically grown foods.

Goodwin Creek Gardens (P.O. Box 83, Williams, OR 97544, 503-846-7357, free catalog). Organically grown seeds for native American herbs.

Halcyon Gardens (P.O. Box 75, Wexford, PA 15090, 412-935-2233, free catalog). Organically grown herb seeds.

Hanna Anderson (1010 NW Flanders, Portland, OR 97209, 800-346-6040, free catalog). Natural fiber clothing for children.

Harmony Farm Supply (P.O. Box 460, Graton, CA 95444, 707-823-9125, free catalog). Organic gardening supplies.

Health Ceramics (400 Gate 5 Road, Sausalito, CA 94965, 415-332-3732, free catalog). Pottery with recycled glaze.

Hearthsong (6519 N. Galena Road, Peoria, IL 61614, 800-325-2502, free catalog). Natural children's art and craft supplies and toys.

Heart of Vermont (The Old Schoolhouse, Route 132, Sharon, VT 05065, 800-639-4123, 802-763-2720, free catalog). Beds, bedding, furniture, personal care products—high standards for environmental integrity.

Hendricksen's Natürlich (8031 Mill Station Road, Sebastopol, CA 95472, 707-829-3959, free catalog). All kinds of natural carpet and flooring materials.

Homespun Fabrics & Draperies (P.O. Box 3223, Ventura, CA 93006, 805-642-8111, free catalog). Unbleached textured cotton draperies and fabrics for making draperies.

Hummer Nature Works, The (Reagan Wells Canyon, P.O. Box 122, Uvalde, TX 78801, 512-232-6167, free catalog). Sustainably harvested cedar products.

Integral Energy Systems (109 Argall Way, Nevada City, CA 95959, 916-265-8441, free catalog). Renewable energy systems.

Jade Mountain (P.O. Box 4616, Boulder, CO 80306, 303-449-6601, catalog $3.00). Renewable energy systems.

Jaffe Brothers (P.O. Box 636, Valley Center, CA 92082–0636, 619-749-1133, free catalog). Basic organic staples in bulk, good prices.

Janice Corporation (198 Route 46, Budd Lake, NJ 07828, 800-526-4237, free catalog). Cotton mattresses and general merchandise particularly for those who are chemically sensitive.

Jantz Design (P.O. Box 3071, Santa Rosa, CA 95402, 800-365-6563, free catalog). Beds and bedding made from organic cotton and wool.

J. L. Hudson, Seedsman (P.O. Box 1058, Redwood City, CA 94064, no phone, free catalog). An impressive collection of open-pollinated seeds from around the world.

Kettle Care (710 Trap Road, Columbia Falls, MT 59912, 406-892-3294, free catalog). Organically grown herbal body care products.

Kinsman Company, The (River Road, Point Pleasant, PA 18950, 215-297-5613, free catalog). Organic gardening supplies.

Lyn-Bar Enterprises (33 Plandome Road, Manhasset, NY 11030, 516-365-6961, free catalog). Custom printing on recycled papers with soy-based ink.

Meadowbrook Herb Garden (Route 138, Wyoming, RI 02898, 401-539-7603, send self-addressed, stamped envelope for catalog). Biodynamically and organically grown herbal products.

Mia Rose Products (1374 Logan, Unit C, Costa Mesa, CA 92626, 714-662-5465, free catalog). Natural air freshener.

Motherwear (P.O. Box 114, Northampton, MA 01061, 413-586-3488, free catalog). Products for natural mothering.

Native Seeds/SEARCH (2509 N. Campbell Avenue, #325, Tucson, AZ 85719, 602-327-9123, free catalog). The catalog of a nonprofit seed-conservation organization working to preserve the traditional native crops of the Southwest. Over 200 varieties.

Natural Baby Company (816 Silvia Street, 800B, Trenton, NJ 08628, 800-388-BABY, free catalog). Natural baby products.

Natural Choice (1365 Rufina Circle, Santa Fe, NM 87501, 800-621-2591, free catalog). Importer of Livos natural paints and other household products.

Natural Gardening Company, The (217 San Anselmo Avenue, San Anselmo, CA 94960, 415-456-5060, free catalog). Organic gardening supplies.

Natural Pet Care Company (8050 Lake City Way, Seattle, WA 98115, 800-962-8266, free catalog). Natural pet care products.

Nature's Colors (424 LaVerne Avenue, Mill Valley, CA 94941, 415-388-6101, send large self-addressed envelope with three stamps for catalog). Natural cosmetic products.

Necessary Trading Company (P.O. Box 305, New Castle, VA 24127, 703-864-5103, free catalog). Organic gardening supplies.

New Cycle (P.O. Box 1775, Sebastopol, CA 95473, 707-829-2744, free catalog). Women's menstrual products made from organic cotton.

Nigra Enterprises (5699 Kanan Road, Agoura, CA 91301, 818-889-6877, free consultation by phone). Air and water purification devices.

Nokomis Farms (3293 Main Street, East Troy, WI 53120, 414-642-9665, free catalog). Bread made from biodynamically grown wheat.

Ocarina Textiles (P.O. Box 65, Old Mystic, CT 06372, 203-536-7480, write for fabric swatch prices). Handmade organic cotton fabrics.

Organic Cottons (104 Clearfield Lane, Coatesville, PA 19320, 215-383-7774, free catalog). Discount prices on organically grown cotton clothing for the whole family.

Patagonia (P.O. Box 8900, Bozeman, MT 59715, 800-638-6464, free catalog). Natural and recycled fabric clothing.

Peaceful Valley Farm Supply (P.O. Box 2209, Grass Valley, CA 95945, 916-272-4769, free catalog). Organic gardening supplies.

Photocomm (930 Idaho Maryland Road, Grass Valley, CA 95945, 800-544-6466, free catalog). Energy-saving products and renewable energy systems.

Pueblo to People (P.O. Box 2545, Houston, TX 77252-2545, 800-843-5257, free catalog). Features colorful clothing, toys, jewelry, furniture, and other items made by cooperatives in nine Central and South American countries.

Quill (P.O. Box 4700, Lincolnshire, IL 60197-4700, 708-634-4800, free catalog). Discount prices on bulk purchase of recycled office paper products.

Real Goods (966 Mazzoni Street, Ukiah, CA 95482-3471, 800-762-7325, free catalog). Biggest selection of energy-saving and renewable energy products, and general merchandise.

Sajama Alpaca (P.O. Box 1209, Ashland, OR 97520, 800-736-0949, write for sample prices). Naturally dyed wool yarns.

Scientific Glass (113 Phoenix NW, Albuquerque, NM 87125, 505-345-7321, free catalog). Water distiller.

Seeds of Change (621 Old Santa Fe Trail, #10, Santa Fe, NM 87501, 505-983-8956, free catalog). This seed bank contains over 6,500 native varieties. The catalog lists hundreds of open-pollinated and organically grown varieties.

SelfCare Catalog (5850 Shellmound Avenue, #390, Emeryville, CA 94608, 800-345-3371, free catalog). Energy-efficient cookware, sun protection devices, and other products to care for your health.

Seventh Generation (Colchester, VT 05446–1672, 800-456-1177, free catalog). Clothing and general household merchandise.

Shelburne Farms (Shelburne, VT 05482, 802-985-8686, free catalog). Cheese made from organic milk.

Simmons Handcrafts (42295 Highway 36, Bridgeville, CA 95526, 800-428-0412, free catalog). Handmade soaps and other personal care items (plus a little general merchandise).

Sinan Company (P.O. Box 857, Davis, CA 95617-0857, 916-753-3104, free catalog). Importer of Auro natural paints and other household products.

Smith & Hawken (25 Corte Madera, Mill Valley, CA 94941, 800-776-3336, free catalog). Smart Wood certified garden furniture and organic gardening supplies.

Talavaya Seeds (P.O. Box 707, Santa Cruz Station, Santa Cruz, NM 87507, 505-753-5801, free catalog). Catalog of a seed bank with more than 1,300 strains of open-pollinated seeds.

Tomato Seed Company, The (P.O. Box 323, Metuchen, NJ 08840, 503-895-2957, free catalog). Organically grown and open-pollinated tomato seeds.

Traditional Products Company (P.O. Box 564, Creswell, OR 97426, no phone, free catalog). Wooden toothbrushes, hairbrushes, and combs.

Tree-Free Eco Paper (121 SW Salmon, #1100, Portland, OR 97204, 800-775-0225, free catalog). A variety of paper products made from hemp and straw.

Vermont Country Store (P.O. Box 3000, Manchester Center, VT 05255–3000, 802-362-4647, free catalog). A variety of natural, durable country goods.

Vreseis (6272 Cedar Street, Santa Susana, CA 93063, 805-522-5381, write for current catalog/samples price). **FoxFibre Colorganics** fabrics and yarns.

Walnut Acres (Penns Creek, PA 17862, 800-433-3998, 717-837-0601, free catalog). Large, full-color catalog of organic foods, including many products made and canned on its farm (cookies, soups, etc.).

Weleda (P.O. Box 769, Spring Valley, NY 10977, 914-352-6145, free catalog). Biodynamically grown herbal cosmetic products.

Williams-Sonoma (P.O. Box 7456, San Francisco, CA 94120–7456, 800-541-1262, free catalog). Clay cookware and other cooking supplies.

Directory of Product Manufacturers

AFM coating products (AFM Enterprises, Riverside, CA)
Air-Crete Foamed-in-Place Insulation (Air-Crete, Weedsport, NY)
Air Therapy air freshener (***Mia Rose Products,*** Costa Mesa, CA)
Alexandra Avery cosmetic products (***Alexandra Avery,*** Birkenfeld, OR)
American Natural Pencils (FaberCastell, Parsippany, NJ)
Arm & Hammer Baking Soda (Church & Dwight, Princeton, NJ), for a handy wheel that gives proper formulas for 37 cleaning uses around the home, write Church & Dwight, 469 N. Harrison Street, Princeton, NJ 08543–5297
Auro paints and household products (imported to America by ***Sinan Company***)
Aware, C.A.R.E., Envirocare, Enviroquest, Forever Green, Green Meadow, New Day's Choice, Project Green, Safe bath and kitchen paper products (Statler Tissue, Medford, MA)
Barbecubes Natural Fruitwood Briquets and Hexaflame MasterStart (WorldWise, San Rafael, CA)
Best Buy Trash Bags and **Full Circle Trash Bags** (Dyna-Pak Corporation, Lawrenceburg, TN)
Bon Ami Cleaning Powder, Cleaning Cake, and **Polishing Cleanser** (Faultless Starch/Bon Ami, Kansas City, MO)
Brooks Pearwood Toothbrush (***Traditional Products Company,*** Creswell, OR)
Capri, Gayety, Gentle Touch, Nature's Choice, Pert bath and kitchen paper products (Pope & Talbot, Portland, OR)
Clear Magic Cleaners (Blue Coral, Cleveland, OH)
Crane papers (Crane & Company, Dalton, MA)
Cycle II recycled paint (Major Paint Company, Torrance, CA)
Descale-it Bathroom Cleaner and **Lime-Eater Bath & Kitchen Cleaner** (Descale-it Company, Tucson, AZ)
Dr. Bronner's Pure Castile Soaps (All-One-God-Faith, Escondido, CA)
Dr. Hauschka cosmetic products (Dr. Hauschka Cosmetics, Wyoming, RI)
Drain Care (Enforcer Products, Cartersville, GA)
Drain King (GT Water Products, Moorpark, CA)
Durotherm Plus cookware (imported by Swiss Kitchen, Greenbrae, CA)
EarthRite Household Cleaners (Benckiser Consumer Products, Danbury, CT)
Eco Fibre Canada (Niagara-on-the-Lake, Ont., Canada)
Ecollection clothing (Esprit, San Francisco, CA)
Eco Sneaks and **Envirolite** shoes (Deja, Inc., Tigard, OR)
Ecos Paint (***Ecos Paint Company,*** Saint Johnsbury, VT)
EcoWorks Ecological Lightbulb (Eco Works, Baltimore, MD)

Eliminate Shower Tub & Tile Cleaner (H.E.R.C., Phoenix, AZ)
Energy Saver computer (IBM, Armonk, NY)
First Alert Fire Extinguishers (BRK Electronics, Aurora, IL)
FoxFibre (Natural Cotton Colours, Wickenburg, AZ)
Fox River recycled printing and writing papers—**Circa '83, Circa Select, Confetti, Early American, Fox River Bond** (Fox River Paper Company, Appleton, WI)
Frigidaire refrigerators (Frigidaire Refrigerator Products, Greenville, MI)
Gehry Collection, The (The Knoll Group, New York, NY)
General Electric light bulbs (General Electric, Fairfield, CT)
Green Forest bath and kitchen paper products (Fort Howard, Green Bay, WI)
Green Paint (Green Paint Company, Manchaug, MA)
Heinz All Natural Cleaning Vinegar (Heinz USA, Pittsburgh, PA), for tips on using vinegar to clean, write Heinz USA, P.O. Box 57, Pittsburgh, PA 15230, attention: Heinz All Natural Cleaning Vinegar
Image Carpets (800-722-2504)
J. R.'s All Purpose Cleaner Degreaser and **Tub & Shower Cleaner** (Eden Sales and Marketing, Phoenix, AZ)
Kiss My Face Lipstick (Kiss My Face, Gardiner, NY)
Lights of America light bulbs (Lights of America, Walnut, CA)
Livos paints and household products (***Natural Choice,*** Santa Fe, NM)
Manville Building Insulation (Manville Building Insulation, Denver, CO)
Mohawk printing and writing papers (Mohawk Paper Mills, Cohoes, NY)
Naturally Free (***Bug Off,*** Lexington, VA)
Nature's Colors cosmetics (***Nature's Colors,*** Mill Valley, CA)
New Cycle women's menstrual products (***New Cycle,*** Sebastopol, CA)
New World bed and bath linens (Fieldcrest Cannon, Eden, NC)
O wear clothing (O wear, Greensboro, NC)
Planet Cleaning Products (Planet Products, Victoria, B.C., Canada)
Proterra printing and writing papers (Georgia-Pacific, Atlanta, GA)
Rain Crystal water distiller (***Scientific Glass,*** Albuquerque, NM)
Renew bath and kitchen paper products (Webster Industries, Peabody, MA)
Römertopf clay cookware (Reco International, Port Washington, NY)
Save the Earth Bed Pillow (Hollander, Boca Raton, FL)
Seagull IV water filter (General Ecology, Lionville, PA)
Simply Cotton bed and bath linens (J. P. Stevens, New York, NY)
Speick shaving products (Walter Rau Gmbh & Company, Germany)
Spred 2000 and **Lifemaster 2000** (The Glidden Company, Huron, OH)
Strathmore papers (Strathmore Paper Company, Westfield, MA)
Sucanat Organic Sugar (Pronatec International, Peterborough, NH)
Sylvania light bulbs (GTE Products Corporation, Winchester, KY)
Today's Choice bath and kitchen paper products (Confab Company, Mission Viejo, CA)
Tom's personal care products (Tom's of Maine, Kennebunk, ME)
Tree-Free bath and kitchen paper products (Tree-Free, Midford, MA)
Tree-Free Eco Paper (***Tree-Free Eco Paper,*** Portland, OR)
Water Safe WS/RO5 water filter (***Coast Filtration,*** Brea, CA)
Weleda cosmetic products (***Weleda,*** Spring Valley, NY)
WorldWise Easy-Clear Sink Trap (WorldWise, San Rafael, CA)

Directory of Organizations

Aluminum Recycling Association (900 19th Street NW, Washington, DC 20006, 202-862-5100). Offers educational materials on aluminum and recycling.

American Forest and Paper Association (1111 19th Street NW, Washington, DC 20036, 800-878-8878). Offers educational materials on recycling and the paper and forest products industries.

Americans for Safe Food (1875 Connecticut Avenue NW, Suite 300, Washington, DC 20009-5728, 202-332-9110). A coalition of more than 80 safer-food groups. Publishes list of organic food sources.

Arts and Crafts Materials Institute (715 Boyleston Street, Boston, MA 02116, 617-266-6800). Certifies children's and adults' art materials with "CP Nontoxic," "AP Nontoxic," and "Health Label" seals. For a list of approved products, send a self-addressed stamped envelope. Also publishes a booklet called *What You Need to Know About the Safety of Art and Craft Materials.*

Bio-Dynamic Farming and Gardening Association (P.O. Box 550, Kimberton, PA 19442, 215-935-7797). Founded in 1938 this is the oldest group advocating an ecological, sustainable approach to agriculture. Publishes a newsletter and has books, advisory service, training programs, and conferences on biodynamic growing.

Bio-Integral Resource Center (P.O. Box 7414, Berkeley, CA 94707, 415-524-2467). The very best source for practical information on the least toxic methods for managing pests. Members can get advice by mail or phone for virtually any pest problem. Publishes a quarterly newsletter and booklets on almost every pest you'll find in your house.

California Organic Wine Alliance (7740 Fairplay Road, Somerset, CA 95684, 209-245-3248). Has lists of organically grown wines.

Center for Occupational Hazards (5 Beekman Street, Suite 1030, New York, NY 10038, 212-227-6220). A national clearinghouse on hazards in the arts. Answers written and telephone inquiries, and distributes more than 70 books, pamphlets, articles, and data sheets.

Citizen's Clearinghouse for Hazardous Waste (P.O. Box 6806, Falls Church, VA 22040, 703-237-2249). A service center for grass-roots environmental groups. Helps with community toxics issues such as hazardous waste dumps and other toxic exposures.

Coalition for Environmentally Responsible Economies (711 Atlantic Avenue, Boston, MA 02111, 617-451-0927). A coalition of social investors, major environmental groups, public pensions, labor organizations, and public interest groups working to create a healthy economy and a healthy environment.

Committee for Sustainable Agriculture (P.O. Box 1300, Colfax, CA 95713, 916-346-2777). An educational organization to provide information on ecological farming to farmers to help them make the change. Publications are also relevant to the general public.

Community Supported Agriculture of North America (c/o WTIG, 818 Connecticut Avenue NW #800, Washington, DC 20006, 202-785-5135). Has an annotated directory of CSAs across the country, and a book on how to start a CSA.

Conservation and Renewable Energy Inquiry and Referral Service (P.O. Box 8900, Silver Spring, MD 20904, 800-523-2929). Provides basic information on renewable energy technologies and energy conservation. Operated by the U.S. Department of Energy.

Coop America (2100 M Street NW, Suite 310, Washington, DC 20063, 800-424-2667, 202-872-5307). Works toward environmental quality, workplace democracy, and a peaceful society. In addition to a mail-order catalog, they also have a Directory of Responsible Businesses, insurance plans that invest premiums in socially responsible ways, a Responsible Travel Agency, access to investment funds, and a quarterly magazine.

Council for Textile Recycling (Suite 1212, 7910 Woodmont Avenue, Bethesda, MD 20814, 301-718-0671). Has information on recycling textiles.

Council on Economic Priorities (30 Irving Place, New York, NY 10003, 212-420-1133). Has information on social responsibility of various corporations and sponsors activities that encourage corporations to be good citizens.

Cultural Survival (215 First Street, Cambridge, MA 02142, 617-621-3818). Has a program to market sustainably harvested rain-forest products as part of their work advocating the rights of indigenous peoples and ethnic minorities worldwide. Many commercial products obtain their raw rain-forest materials through Cultural Survival programs. They also sell plain Brazil nuts, cashews, and cashew fruit.

Demeter Association for Certification of Bio-Dynamic Agriculture (4214 National Avenue, Burbank, CA 91505, 818-843-5521). Certifies biodynamic food.

EPA Office of Air Quality and Standards (MD13 RTP NC 27711—this really is the address they gave me and it works!). Information liaison for LCA study. The project was designed to develop a methodology for evaluating the environmental and public health consequences of consumer products throughout their life cycles and to develop a mechanism for providing this information to consumers. As of 1993 the focus of the project was to discover and forge consensus between those who already use some form of LCA. They were well on their way in developing a "multi-media, upstream-downstream, holistic environmental assessment tool" and were preparing papers on inventory guidelines and principles, impact analysis, data quality and availability, and applications of LCA.

Farm Verified Organic (Box 45, Redding, CT 06875, 203-544-9896). Certifies organically grown food.

Food and Water (225 Lafayette Street, Suite 612, New York, NY 10012, 800-EAT-SAFE). Works to educate the public on food irradiation hazards.

Forest Stewardship Council (R.R. 1, Box 188, Richmond, VT, 802-434-3101). Developing program to accredit wood certification programs worldwide.

Glass Packaging Institute (1801 K Street NW, Suite 1105-L, Washington, DC 20006, 202-887-4850). Offers educational materials on recycling glass.

Green Seal (1250 23rd Street NW, Suite 275, Washington, DC 20037-1101, 202-331-7337). Grants environmental seal of approval and sets standards for choosing products. Standards are available to consumers.

Household Hazardous Waste Project (P.O. Box 108, Springfield, MO 65804, 417-836-5777). A community health and environment education program sponsored by the Missouri Department of Natural Resources. Publishes the highly informative and comprehensive *Guide to Hazardous Products Around the Home.* which discusses in detail everything from labeling to hazardous ingredients and homemade alternatives.

INFORM (381 Park Avenue South, New York, NY 10016, 212-689-4040). Publishes a number of books on toxics, including *Tackling Toxics in Everyday Products,* a directory of organizations concerned about toxics in consumer and building products.

Institute For Sustainable Forestry (P.O. Box 1580, Redway, CA 95560, 707-923-4719). Administers Pacific Certified Ecological Forest Products certification program.

Massachusetts Audubon Society (South Great Road, Lincoln, MA 01773, 617-259-9500). Offers eight booklets that describe energy solutions clearly and simply, including weatherization, solar, insulation, financing tips, and more.

Mothers and Others for Pesticide Limits (a project of the Natural Resources Defense Council, 40 West 20th Street, New York, NY 10011, 212-727-4474). A national organization that presses for reforms in pesticide regulations to ensure the availability of safe produce. Publishes *For Our Kids' Sake: How to Protect Your Child Against Pesticides in Food.*

National Appropriate Technology Assistance Service (P.O. Box 2525, Butte, MT 59702-2525, 800-428-2525, 800-428-1718). Toll-free service answers technical questions on wood stoves, insulation, wind, furnaces, appliances, water heaters, and renewable energy economics.

National Coalition Against the Misuse of Pesticides (530 Seventh Street SE, Washington, DC 20003, 202-534-5450). An organization of groups and individuals working to better protect the public from toxic pesticides. Has information on pesticide dangers and alternatives.

National Recycling Coalition (1101 30th Street NW, Suite 305, Washington, DC 20007, 202-625-6406). Offers fact sheets, reports, and directories related to recycling.

National Toxics Campaign (37 Temple Place, 4th Floor, Boston, MA 02111, 617-482-1477). Works to implement citizen-based preventative solutions to the nation's toxic problems. Publishes newsletter and manuals on toxic protection and pollution prevention.

Natural Organic Farmers Association (RFD No. 2, Barre, MA 01005, 508-355-2853). Certifies organically-grown food.

Northwest Coalition for Alternatives to Pesticides (P.O. Box 1393, Eugene, OR 97440, 503-344-5044). Provides information packets on pesticide issues and fact sheets on pesticides commonly used.

NSF International/Canadian Standards Association (P.O. Box 130140, Ann Arbor, MI 48113-0140, 313-769-8010). Is coordinating an effort to come to agreement on standards between all interested groups, and has written guidelines for LCA.

Nuclear Free America (325 E. 25th Street, Baltimore, MD 21218, 301-235-3575). International clearinghouse and resource center for Nuclear Free Zones that tracks products made by nuclear weapons contractors and their subsidiaries.

Organic Crop Improvement Association (3185 Township Road 179, Bellefontaine, OH 43311, 513-592-4983). Certifies organically grown food.

Organic Foods Production Association of North America (P.O. Box 31, Belchertown, MA 01007, 413-332-6821). Certifies organically grown food.

Organic Grapes into Wine Alliance (54 Genoa Place, San Francisco, CA 94133, 800-477-0167). Educates consumers and promotes organic grapes and wines.

Public Citizen (215 Pennsylvania Avenue SE, Washington, DC 20003, 202-546-4996). Founded in 1971 by Ralph Nader, it publishes the *National Directory of U.S. Energy Periodicals* (includes more than 700 publications on all kinds of energy) and the *National Directory of Safe Energy Organizations* (more than 1,000 citizen and other nonprofit groups actively promoting energy efficiency and renewable technologies, or opposing nuclear energy).

Rachel Carson Council (8940 Jones Mill Road, Chevy Chase, MD 20815, 301-652-1877). Maintains an extensive library on chemical contamination and its health impact.

Rainforest Action Network (301 Broadway, Suite A, San Francisco, CA 94133, 415-398-4404). They publish *The Wood Users Guide* by Pamela Wellner and Eugene Dickey which discusses the problems with rain-forest woods and gives a comprehensive overview of domestic wood and nonwood alternatives. Lists tree species and descriptions, and suppliers of ecologically sensitive woods. They also have a small catalog of books on rain-forest issues as well as a few products that contain rain-forest ingredients.

Rainforest Alliance (270 Lafayette, Suite 512, New York, NY 10012, 212-677-1900). Administers Smart Wood certification program for rain-forest woods.

Remote Access Chemical Hazards Electronic Library (RACHEL, Environmental Research Foundation, P.O. Box 3541, Princeton, NJ 08543-3541, 609-683-0707). A computer database of information about hazardous materials, including health and some environmental effects.

Rocky Mountain Institute (1739 Snowmass Creek Road, Snowmass, CO 81654-9199, 303-927-3128). A leader in energy resource-efficiency research and policy. Publications include *The Resource-Efficient Housing Guide* and a newsletter.

Scientific Certification Systems (1611 Telegraph Avenue, Suite 1111, Oakland, CA 94612-2113, 510-832-1415). Grants independent certification of environmental claims.

Society of the Plastics Industry (1275 K Street NW, Suite 400, Washington, DC, 20005, 202-371-5200). Has information on plastics recycling.

Solid Waste Alternatives Project—Environmental Action Foundation (1525 New Hampshire Avenue NW, Washington, DC 20036). Has information on recycling.

Solid Waste Composting Council (114 South Pitt Street, Alexandria, VA 22314, 703-739-2401). The only association devoted to promoting composting. Developing national standards for municipal and home compost.

Steel Can Recycling Institute (680 Andersen Drive, Pittsburgh, PA 15220, 412-922-2772, 800-876-SCRI). Has information on recycling steel cans.

Texas Organic Cotton Growers Association (201 West Broadway, Brownfield, TX 79316, 806-637-4547). An organization of growers and others that promotes organic cotton.

Toxics Release Inventory. To find the database nearest you, call the EPA hotline at 800-535-0202.

Washington Toxics Coalition (4516 University Way NE, Seattle, WA 98105, 206-632-1545). Works to reduce our society's reliance on toxic chemicals by acting as a public information clearinghouse and providing education to individuals, government agencies, and lawmakers. Has information on toxics and alternatives and sponsors a Home Safe Home project that identifies and promotes alternatives to toxic household products.

Woodworkers Alliance for Rainforest Protection (Box 133, Coos Bay, OR 97420), a grassroots association of woodworkers and wood users that supports sustainable development of forest resources and provides information about responsible timber management and wood use.

WorldWatch Institute (1776 Massachusetts Avenue NW, Washington, DC 20036, 202-452-1999). "Tracks key indicators of the Earth's well-being" and disseminates information on our progress toward a sustainable world. Publishes ***WorldWatch*** magazine and many booklets on related topics.

Index